Inorganic Chemistry

FOR

DUMMIES

A Wiley Brand

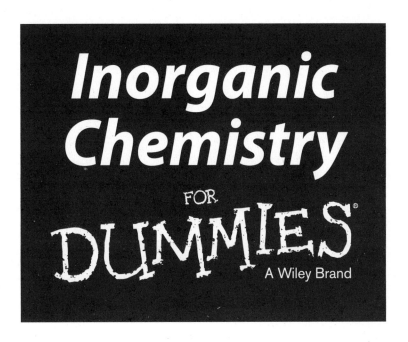

Inorganic Chemistry

FOR

DUMMIES®

A Wiley Brand

by Michael L. Matson and Alvin W. Orbaek

FOR

DUMMIES®

A Wiley Brand

Inorganic Chemistry For Dummies®

Published by
John Wiley & Sons, Inc.
111 River St.
Hoboken, NJ 07030-5774
www.wiley.com

Copyright © 2013 by John Wiley & Sons, Inc., Hoboken, New Jersey

Published by John Wiley & Sons, Inc., Hoboken, New Jersey

Published simultaneously in Canada

For general information on our other products and services, please contact our Customer Care Department within the U.S. at 877-762-2974, outside the U.S. at 317-572-3993, or fax 317-572-4002.

For technical support, please visit www.wiley.com/techsupport.

Wiley also publishes its books in a variety of electronic formats and by print-on-demand. Some content that appears in standard print versions of this book may not be available in other formats. For more information about Wiley products, visit us at www.wiley.com.

Library of Congress Control Number: 2013932110

ISBN 978-1-118-21794-8 (pbk); ISBN 978-1-118-22882-1 (ebk); ISBN 978-1-118-22891-3 (ebk); ISBN 978-1-118-22894-4 (ebk)

Manufactured in the United States of America

10 9 8 7 6 5 4 3 2 1

About the Authors

Michael L. Matson started studying chemistry at the U.S. Naval Academy in Annapolis, Maryland. After leaving the Navy, Michael started a PhD program at Rice University, studying the use of carbon nanotubes for medical diagnosis and treatment of cancer. Specifically, Michael focused on internalizing radioactive metal ions within carbon nanotubes: Some radioactive metals could be pictured with special cameras for diagnosis, whereas others were so powerful they could kill cells for treatment. It was at Rice that Michael and Alvin met. Following Rice, Michael went to the University of Houston-Downtown to begin a tenure-track professorship. Happily married to a woman he first met in seventh grade, Michael has two young children, a yellow Labrador retriever named Flounder, is a volunteer firefighter and sommelier, and enjoys CrossFitting.

Alvin W. Orbaek was introduced to chemistry at Rice University (Houston, Texas) by way of nanotechnology, where he studied single-walled carbon nanotubes, transition metal catalysts, and silver nanoparticles. He had previously received a degree in Experimental Physics from N.U.I. Galway (Ireland) and moved into the study of space science and technology at the International Space University (Strasbourg, France). He received a position on Galactic Suite, an orbiting space hotel. To date, he enjoys life by sailing, snowboarding, and DJing. He has been spinning vinyl records since the Atlantic Hotel used to rave, and the sun would set in Ibiza. He hopes to empower people through education and technology, to that effect he is currently completing a PhD in Chemistry at Rice University.

Dedications

Michael: To my wife, Samantha.

Alvin: To Declan, Ann Gitte, Anton, Anna-livia, and Bedstemor.

Authors' Acknowledgments

Michael: I'd like to acknowledge the immeasurable amounts of assistance from Matt Wagner, Susan Hobbs, Lindsay Lefevere, Alecia Spooner, and Joan Freedman.

Alvin: Without John Wiley & Sons, there would be no book, and for that I am very grateful. Particularly because of the very positive and professional attitude by which they carry out their business; thanks for getting it done. It was a blessing to work with you. In particular, I would like to mention Alecia Spooner, Susan Hobbs (Suz), and Lindsay Lefevere, and thanks to the technical editors (Reynaldo Barreto and Bradley Fahlman) for their crucial input. I would also like to thank Matt Wagner for invaluable support and assistance. And to Mike Matson, thank you for the invitation to write this book.

I have had many teachers, mentors, and advisors throughout the years, but there are five who deserve attention. Andrew Smith at Coleenbridge Steiner school, where I enjoyed learning a great deal. John Treacy, who made every science class the most riveting class each day. Pat Sweeney, whose habit of teaching would leave anyone engrossed in mathematics. To Ignasi Casanova for his mentorship and introduction to the nanos. And Andrew Barron, both my PhD advisor and mentor, to whom I owe a great deal of credit, due in no small part to his measure of tutelage.

But all this stands upon a firm foundation that is based on the support of Dec, Gitte, Anton, and Anna; here's to next Christmas — whenever. There are many other friends and family who have contributed to this work, too many to mention them all. But I'd especially like to thank my colleagues from the Irish house, who so graciously agreed to read through the text, namely Alan Taylor, Nigel Alley, and Stuart Corr. Also to Sophia Phounsavath and Brandon Cisneros for proofreading. Jorge Fallas for the Schrödinger equation. To Gordon Tomas for continued support of my writing. And to Gabrielle Novello, who fed me wholesome foods while I otherwise converted coffee and sleepless nights into this book. And to Valhalla for those nights when work was not working for me. And to PHlert, the best sailing program on this planet, or any other.

Publisher's Acknowledgments

We're proud of this book; please send us your comments at http://dummies.custhelp.com. For other comments, please contact our Customer Care Department within the U.S. at 877-762-2974, outside the U.S. at 317-572-3993, or fax 317-572-4002.

Some of the people who helped bring this book to market include the following:

Acquisitions, Editorial, and Media Development

Project Editor: Susan Hobbs

Acquisitions Editor: Lindsay Lefevere

Copy Editor: Susan Hobbs

Assistant Editor: David Lutton

Editorial Program Coordinator: Joe Niesen

Technical Editors: Reynaldo Barreto, Bradley Fahlman

Editorial Manager: Carmen Krikorian

Editorial Assistant: Rachelle Amick

Art Coordinator: Alicia B. South

Cover Photo: © Laguna Design / Science Source

Cartoons: Rich Tennant (www.the5thwave.com)

Composition Services

Project Coordinator: Sheree Montgomery

Layout and Graphics: Carrie A. Cesavice, Joyce Haughey, Brent Savage

Proofreaders: Lindsay Amones, John Greenough, Jessica Kramer

Indexer: BIM Indexing & Proofreading Services

Publishing and Editorial for Consumer Dummies

  Kathleen Nebenhaus, Vice President and Executive Publisher

Publishing for Technology Dummies

  Andy Cummings, Vice President and Publisher

Composition Services

  Debbie Stailey, Director of Composition Services

Contents at a Glance

Table of Contents

Introduction

● ●

*I*norganic chemistry deals with all the atoms on the periodic table, the various rules that govern how they look, and how they interact. At first glance, trying to understand the differences among 112 atoms might seem like a mammoth task. But because of the periodic table, we can bunch them up into groups and periods and make them much easier to grasp.

So welcome to *Inorganic Chemistry For Dummies*. We hope that through this book you come to learn a great deal about the environment around you, what materials you use on a regular basis, and why some materials are more important to us than others. This book is fun and informative, while at the same time insightful and descriptive. And it's designed to make this fascinating and practical science accessible to anyone, from the novice chemist to the mad scientist.

About This Book

This book was written in such a way that you can start in any chapter you choose, in the chapter that interests you the most, without having to read all the chapters before it. But the chapters build on material from one chapter to the next, so if you feel more background would help you, feel free to start with Chapter 1. You can also make use of the numerous cross references in each chapter to find pertinent information. But it can also be read like a study guide to help a student understand some of the more complicated aspect of this fascinating science.

We tried to make the information as accessible as possible. Each chapter is broken down into bite-sized chunks that make it easy for you to quickly digest and understand the material presented. Some of the chunks are further broken down into subsections when there's special need to elaborate further on the concepts being discussed.

Science is a process that requires lots of imagination. It requires more imagination than memory, especially as you start to learn more and more about a certain topic. To help with your imagination we have tried to include helpful graphics and artwork that complement the writing within the text. Further to this we include many real-world examples and interesting historical or scientific tidbits to keep your curiosity piqued.

Conventions Used in This Book

Science progressed more rapidly in the last 200 years than it had in the few thousand years previous. A great deal of this success came from the agreement among scientist to create and use a set of standard conventions. The two most important conventions are the periodic table and the international system of units, called SI units. SI units are based on the metric system, and it's more common to see temperature expressed as Celsius than Fahrenheit. And you see lengths expressed in meters instead of inches and feet. Weights and mass are expressed in terms of grams instead of pounds or stone.

And the following conventions throughout this text make everything consistent and easy to understand:

- ✔ All Web addresses appear in `monofont`.
- ✔ New and key terms appear in *italics* and are closely followed by an easy-to-understand definition.
- ✔ **Bold** text highlights the action part of numbered steps.

What You Don't Need to Read

Sidebars are highlighted in gray-shaded boxes so they're easy to pick out. They contain fun facts and curious asides, but none of their information is crucial to your understanding of inorganic chemistry. Feel free to just skip over them if you prefer.

Foolish Assumptions

As authors of *Inorganic Chemistry For Dummies* we may have made a few foolish assumptions about the readership. We assume that you have very little background in chemistry, and possibly none at all; that you're new to inorganic chemistry, and maybe you have never heard of the subject before. We assume that you know what chemistry is, but not much more than that. This book begins with all the general chemistry info that you need to grasp the concepts and material in the rest of the book. If you have some understanding of general chemistry, however, all the better.

You may be a medical student who needs to brush up in inorganic chemistry, or a high school student getting ready for a science fair, or even a freshman or junior at college. We've tailored this book to meet all your needs, and we sincerely hope you find great explanations about the concepts presented that are also engaging, interesting, and useful.

When you finish reading this book and your interest in chemistry is heightened, we recommend that you go to a local bookseller (second-hand book stores are a personal favorite) and find more books that offer other perspectives on inorganic chemistry. There are also excellent resources on the Internet, and many schools make class notes available online. But the best way to get involved in chemistry is by doing it. Chemistry is a fun and exciting field, made evident when you conduct chemistry experiments. Keep an eye out for demonstration kits that enable you to do your own experiments at home. And note that the last chapter of this book offers ten really cool experiments, too.

How This Book Is Organized

This book is organized into multiple parts that group topics together in the most logical way possible. Here's a brief description of each section of *Inorganic Chemistry For Dummies*:

Part I: Reviewing Some General Chemistry

Here you are introduced to science in general, and we give you the basic tenets of general chemistry that help you throughout the rest of the book.

In Chapter 1, you start with an introduction to inorganic chemistry, what it is, and why it is important. You learn how it's different from organic chemistry and how this difference is important for technology and society.

The following chapters of this section deal with topics that are covered in many general chemistry textbooks, but these chapters cover the topics in greater detail than a general chemistry textbook. In Chapter 2 we explain what the atom looks like, how it's structured, and why this is important for inorganic chemistry. In particular, this chapter delves into the periodic table and how the structure of the atom is described. Chapter 3 introduces oxidation and reduction chemistry that helps you understand why many chemical reactions take place. It deals with the electrons that each atom has and how the electrons can be shuttled around from atom to atom. Then in Chapter 4 we focus on the nucleus and how changes to the nucleus lead to nuclear chemistry. And finally we end this section by talking about acid-base chemistry because this can help you understand the many ways in which atoms and molecules interact with one another.

Part II: Rules of Attraction: Chemical Bonding

In this section we talk about the various ways that atoms can bond with one another. In Chapter 6 we introduce covalent bonding. Chapter 7 deals with molecular symmetry, not just for inorganic chemistry but also fundamental to many of the physical sciences. Ionic and metallic bonding are detailed in Chapter 8.

Chapter 9, like all of the chapters, can be read as a standalone chapter, but it's much easier to understand if you read through the three preceding chapters. If you get stuck on coordination complexes, however, refer back to the previous three chapters for a little background information.

Part III: It's Elemental: Dining at the Periodic Table

The periodic table contains over 100 separate and unique elements, which are described in Part III. We cover all the important elements; and to make it easier to digest, we've broken them down into five related chapters. Each chapters deals with elements that are similar to each other, making them easier to understand.

To get the ball rolling we introduce hydrogen in Chapter 10, because it's the most abundant element in the universe and can be found in many chemicals and materials. We then move from left to right on the periodic table, starting off with the alkali and alkali earth elements in Chapter 11. We guide you through the periodic table to the main group elements in Chapter 12, the transition metals in Chapter 13, and finally round out Part III with the lanthanides and actinides in Chapter 14.

Part IV: Special Topics

These chapters cover what makes the study of inorganic chemistry so interesting and also distinguishes it from organic chemistry. However, you will find a great deal of overlap with other fields of study such as material science, physics, and biology.

Inorganic chemistry became a modern science with the advent of organometallic chemistry, described in Chapter 15. Chapter 16 shows you how practical and important catalysis is to the modern world in which we live. Chapter 17 deals with the inorganic chemistry of living systems and the environment. The subject matter makes this chapter unique from the others in this section. This is also true for Chapter 18 where we describe solid state chemistry, the basis of the information technology revolution. Chapter 19 gives you a quick introduction to one of the most interesting and promising technological developments of the modern age, namely nanotechnology.

Part V: The Part of Tens

To make this book even easier to grasp and read, we compiled three important lists to help you in your study of inorganic chemistry. In Chapter 20, we introduce and explain ten common household products. Then, in Chapter 21, you meet ten of the most important Nobel Prizes that were awarded to chemists. Chapter 22 introduces ten instruments and techniques that are commonly found and used in laboratories across the globe. And finally we give you ten experiments that you can try out at home in Chapter 23. Remember, one of the most fun parts of chemistry is doing chemistry, and this chapter gives you some fun experiments to try.

Icons Used in This Book

Throughout this book icons are used to draw your attention to certain information.

This is not often used here, but the Tip icon indicates that some information may be especially useful to you.

When you see the Remember icon you should understand that this information is quite important to understanding the concepts being explained. If you are studying inorganic chemistry, this is one of the most important icons to look for. It can indicate a definition, or be a concise explanation of a concept; at other times it indicates information to help you grasp how various concepts overlap.

The Warning icon tells you to pay close attention to what's being said because it indicates where a potentially dangerous situation may arise.

The Technical Stuff icon is used to indicate detailed information; for some people, it might not be necessary to read or understand.

Where to Go from Here

You might be taking an inorganic chemistry course, or maybe you're just curious about the world around you. Regardless, if you're looking for something specific, you can find it by checking the index or maybe even the glossary. When you know where to find what you are looking for, go right ahead and jump in. And enjoy.

Part I
Reviewing Some General Chemistry

The 5th Wave   By Rich Tennant

"You can take that old jar for your science project,
I'm sure I have some baking soda you can borrow,
and let's see, where's that old particle accelerator
of mine... here it is in the pantry."

In this part . . .

You navigate through some of the basic rules of the road that help guide you as you travel through the science of inorganic chemistry. This starts with a definition of inorganic chemistry and continues with a description of the foundation upon which this subject stands. Inorganic chemistry is the study of all the materials known to humankind, and it includes the study of how all the materials interact with one another.

Chapter 1

Introducing Inorganic Chemistry

*I*norganic chemistry is a practical science. By studying it, you become familiar with the intricate working of processes and materials — from how silicon works in a semiconductor to the reason why steel is stronger than iron. Inorganic chemistry is important for civilization and technological development.

The science of inorganic chemistry covers a great deal of material; in short, it's the chemistry of everything you see around you. Inorganic chemistry explores and defines laws that atoms follow when they interact, including trends in how they react, characteristics they possess, and the materials they make. It may seem daunting at first to think about how many possibilities there are in the science of inorganic chemistry. Fortunately, each new concept builds on another concept in a very logical way.

This chapter explains what to expect when reading this book and should help you find the right section to guide you through your study of inorganic chemistry.

Building the Foundation

Before diving into the particular details of inorganic chemistry, it's helpful to understand some of the prominent ideas in general chemistry that are useful to further appreciate inorganic chemistry.

What difference does it make?

It's important to be able to distinguish between inorganic and organic chemistry. Organic chemistry deals primarily with the reaction of carbon, and its many interactions. But inorganic chemistry deals with all of the other elements (including carbon, too), and it details the various reactions that are possible with each of them. There are a huge number of examples in everyday life that can be described by inorganic chemistry — for example, why metals have so many different colors, or why metal compounds of the same metal can have such varying colors too, like the ones that are used and pigments in paints. It can help to explain how alloys form and what alloys are stronger than others. Or why a dentist uses an acid to open the pores in your teeth before applying an adhesive to make a filling hold fast.

Chemistry is a science of change. It looks at how individual atoms interact with each other and how they are influenced by their environment. We start by explaining what atoms look like, and we describe details of their structure. This is important because the way that the atom is made up determines how reactive that atom is, and as a result of the activity, it can be used by a chemist to make materials. After you have these basics down, you are able to understand the physical properties of many materials based on what atoms they are made from, and why they are made using those specific atoms.

Stemming from this basis of general chemistry we then deal with the specifics of inorganic chemistry. This includes an understanding of approximately 100 atoms that are of practical interest to chemists. To simplify this, inorganic chemistry is understood according to some general trends based on atomic structure that affect the reactivity and bonding of those atoms. This is quite different from the study of organic chemistry that deals with the reactions of just a few atoms, such as carbon, oxygen, nitrogen, and hydrogen. But there is an overlap between inorganic chemistry and organic chemistry in the study of organometallic compounds.

Losing your electrons

In chemical reactions, follow the electrons because electrons hold the key to understanding why reactions take place. Electrons are negatively charged, mobile, and can move from atom to atom; they can be stripped from atoms, too. Atoms are always trying to have just the right amount of electrons to keep stable. If a stable atom has cause to lose or gain an electron, it becomes reactive and starts a chemical process.

The nucleus of an atom has a positive force that attracts electrons. This comes from protons within the nucleus that influence electrons to orbit around the nucleus. As you progress in atomic size, one proton at a time, there is room for one more electron to orbit around the atom.

There are periodic trends that can be seen in the periodic table, the first of which deals with the stability of atoms according to the number of outer electrons in the atom. This is known as *valency*, and it can be used to show why some atoms are more reactive than others. There are many more periodic trends that are associated with the electrons around the atoms, and you can find more examples in Chapter 2.

Take a stable atom, such as iron, for example. Imagine that you remove an electron from iron; it now has a different reactivity. This is known as oxidation chemistry, and it's the focus of Chapter 3. The chemistry of oxidation tracks how electrons are gained or lost from molecules, atoms, or ions. When an electron is lost, the molecule, atom, or ion is said to have an increased oxidation state, or is considered *oxidized*. When the opposite occurs and a molecule, atom, or ion gains an electron, its oxidation state is *reduced*.

Originally named from the common involvement of oxygen molecules in these types of reactions, chemists now realize that oxidation and reduction reactions (sometimes referred to as redox chemistry) can occur among molecules, atoms, and ions without oxygen.

Splitting atoms: Nuclear chemistry

Another area of general chemistry with which you should be familiar is the study of radioactivity, or nuclear chemistry. Specifically, nuclear chemistry deals with the properties of the nucleus of the atoms; that's why it is called nuclear chemistry.

As you progress through the periodic table each successive atom has one more proton and neutron compared with the previous atom. The protons are useful for attracting electrons, and the neutrons are useful for stabilizing the nucleus. When there is an imbalance between the two nuclear particles (proton and neutron), the nucleus becomes unstable, and these types of atoms are called *isotopes*. If they are radioactive, they are called *radioisotopes*, and they can be useful, for example, in medical applications.

Although you may immediately think about nuclear reactors for energy, or nuclear bombs and their incredible devastation, concepts in nuclear chemistry are applied for many other, less dramatic purposes, one such example is carbon dating of ancient materials (see Chapter 4).

The nuclear processes can affect the properties of the atoms, and this can have an effect on the properties of materials that are made with those atoms. For example, there is often a lot of heat generated by radioactive atoms, and this heat can affect material properties. Did you know that much of the potassium in our body is in the form of a radioactive isotope? This accounts for some of the heating within our own bodies (see Chapter 11).

Changing pH

In Chapter 5, we explain the basics of acids and bases, including how the pH scale was developed to quantify the strength of different acids and bases. It's a simple system that ranges in value from pH 1 to pH 14.

Acids have low pH values in the range of pH 1 to pH 7. Bases have high pH values that range from pH 7 to pH 14. In the middle there is pH 7, and this is considered neutral pH, which is also the pH of water. And subsequently is nearly the same pH as blood, demonstrating how important water is to us.

The pH of blood is highly sensitive; if it changes too much, we can get very sick. The preferred range for maintaining stable health is from pH 7.35 to pH 7.45, making blood slightly basic. This simple fact alone highlights the importance of green foods in your diet; they're alkalizing in your body and help maintain a healthy *you*.

Chemists have been working for many years to sort out what specifically makes something an acid or a base. Through this work, multiple definitions of acids and bases have been proposed. As we explain in Chapter 5, there are two important models for examining acid-base chemistry:

- ✔ **Brønsted-Lowry model:** In this model, an acid is a proton (H) donor, whereas a base accepts hydroxyl groups (OH molecule).
- ✔ **Lewis model:** In this model, acids are electron pair acceptors and bases are electron pair donors.

Earlier we said you needed to track the electrons to understand what is happening in various chemical reactions. By using the Lewis model that deals with electron pairs, you can get a good understanding of how reactions occur, by tracking the electron pairs and seeing where they come and go.

It's important to understand the distinction between these two models. The Brønsted-Lowry model was developed when acids and bases were thought to work in aqueous solvents. As a result, it deals only with hydrogen and hydroxyl groups. On the other hand, the Lewis model was developed to show what happens when water isn't the solvent, so it deals with electrons instead.

Getting a Grip on Chemical Bonding

Part II delves into how bonding occurs between atoms, and how to distinguish between the types of bonds that are created. Bonding between atoms is important for all scientists to understand because it affects the properties and applications of materials in profound ways. In practice, there are about

100 atoms that are stable enough to form bonds, but there are only three types of bonding known:

- ✔ **Covalent:** *Covalent bonding* stems from the sharing of electrons and the overlap and sharing of electrons orbitals between atoms. Covalent bonds are very strong as a result of this. Covalent bonds have *directionality*, or a preference for a specific orientation relative to one another, this results in molecules of interesting and specific shapes. As a result, elaborate molecules can be made that have specific structures and symmetry, which we describe in Chapter 7.

- ✔ **Ionic:** *Ionic bonding* occurs when atoms donate or receive electrons rather than share them. One ion is positively charged, and it's balanced by an ion that is negatively charged; they're known as the *cation* and the *anion*, respectively. Each ion is treated as if it's a spherical entity with no distortion of the electron orbital. See more information in Chapter 8.

- ✔ **Metallic:** *Metallic bonds* are similar to ionic bonds, so we describe them both in Chapter 8. The main difference is that in metallic bonds the electrons are shared among all the other atoms in the metal materials. This is known as the delocalization of electrons because they are not found locally around one particular atom. This gives rise to many of the properties of metals.

There aren't strict lines between each type of bond, and sometimes the way atoms bond together is a combination or mixture of more than one bond type. Throughout Part II we explain each of the bond types individually; then in Chapter 9 we will look at how they each influence the formation of molecules known as *coordination complexes,* which include metallic compounds and connecting molecules called *ligands*.

Traveling Across the Periodic Table

There are over 100 known atoms, and it can be overwhelming to try to remember each and every one of them. This is what chemists tried to do before the *periodic table* was created. In Part III, you learn about this important chart that organizes the elements according to their similarities in structure and reactivity. The simplicity and beauty of the periodic table makes it easier to find and compare elements against each other. If the familiar expression "a picture is worth a thousand words" was used to describe inorganic chemistry, then the picture that best describes it is the periodic table. We've devoted the chapters in Part III to exploring the periodic table from one end to the other and describing the key characteristics of each group.

Here you can see what the periodic table looks like. Notice how there are 18 groups from left to right as seen at the top. And there are seven periods going from top to bottom as shown on the left side of the table.

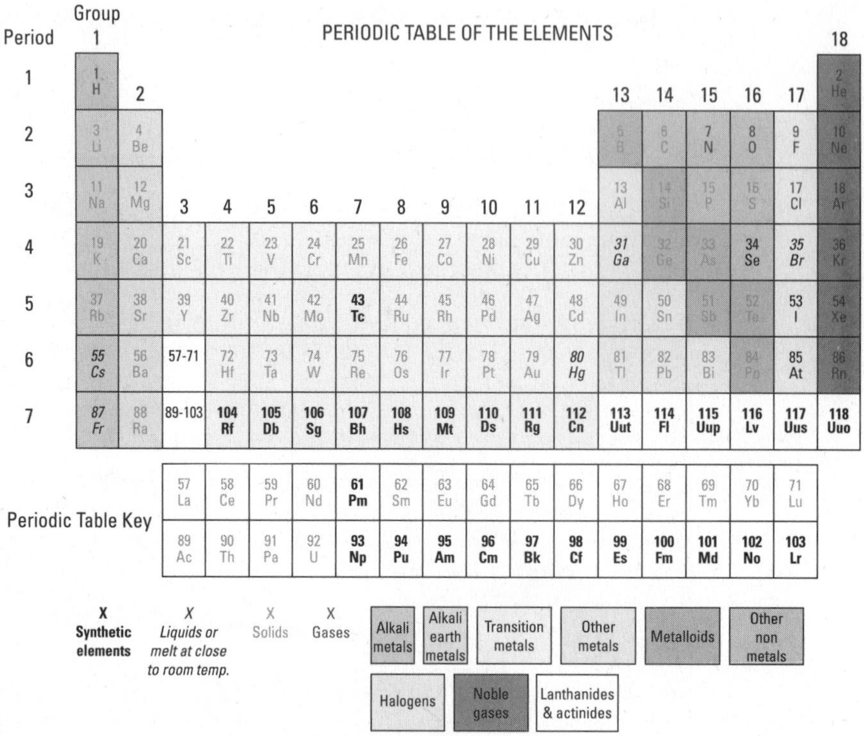

Figure 1-1:
The periodic table of the elements.

Hyping up hydrogen

Hydrogen is one of the most abundant elements in the universe, and Chapter 10 explains the unique and important properties. This element sits at the upper-left corner of the periodic table and serves as the first step in a long line of stepping stones for you to travel across the periodic table. Some points to know about hydrogen include:

- Hydrogen is highly reactive. It lacks one electron in the outer orbital to make it stable, so it has a very reactive valency. This makes it explosive, and for this reason it's usually found as H_2 — two hydrogen atoms bonded together. Because each hydrogen shares the electron, it pacifies the atom.

- Hydrogen is used in a technique called *nuclear magnetic resonance*. This is important because it can be used to elaborate exactly where hydrogen atoms are within a molecule so it can show the structure of the molecule.

- Hydrogen can bond with nearly every single atom on the periodic table, making it a versatile atom.

Moving through the main groups

The most common elements are found in the *main groups* of the periodic table. The main group elements comprise many of the materials we know from everyday experience.

The main group elements include the Group 1 and Group 2 elements on the left side of the table along with Groups 13, 14, 15, 16, 17, and 18 on the right side of the table. The most reactive is on the left side; The most inert and calm reside on the far right. As you might expect, the middle atoms have mixed qualities between these two extremes.

A few of the main group elements have specific qualities recognized by chemists. For example:

- **Alkali and alkaline earth metals:** The elements in the first two columns of the periodic table (excluding hydrogen) are formally known as the alkali and alkaline earth metals, or s-block elements. They are highly reactive and often explosive elements, but also extremely important in biology. Compounds made with Group 1 and 2 elements are often referred to as salts; skip ahead to Chapter 11 to find out why.

- **Noble gases:** The elements in the far right column of the periodic table are the noble gases and are mirror opposites of the alkali and alkaline earth metals. Instead of being reactive, for the most part they are inert, or nonreactive. The noble gases have no need for more electrons, so they generally don't react with other atoms to gain, give, or share electrons. There are some exceptions, however, because the gases of argon, krypton, and xenon can form compounds with fluorine. More of this can be found in Chapter 12.

The rest of the main group elements, called p-block elements, contain the atoms that are associated with life and living matter, including carbon, oxygen, and nitrogen. More information can be found in Chapters 12 and 17.

Transitioning from one side of the table to another

In the center of the periodic table are the elements that transition from the s-block main group elements to the p-block main group elements. These elements are called the *transition metals* or d-block elements. The transition metals act as cushion between the highly reactive elements on the far left and the less reactive elements on the right.

These elements are important for industry and help in the synthesis of organic molecules and medicinal compounds. You can find a number of them in the catalytic converter of your car, for example.

Transitional metals are important because they're used as catalysts in the chemical industry. They're often reactive atoms, and under the appropriate conditions can complete reactions and make large amounts of molecules with a very specific size and shape. Much of the plastic materials that are in use today are made possible on such a grand and industrial scale thanks to the development of catalysis using transition metals. More information about catalysis can be found in Chapter 16. Catalysts make short work of specific chemical reactions; they have the ability to create a product faster, and with less energy.

Some of the reactions that take place in the body do so because of transition metals. For example, the oxygen that we breathe is carried around the body using a compound that has iron at the center. This is called hemoglobin. But the other transition metals can play important roles in the body also, for more information see Chapter 17.

Many transition metals are used in everyday materials that we use regularly. These metals often have interesting electronic and magnetic properties, and because of this they're commonly used in electronic devices. But at the nanoscale (that being the very small scale), they have some other very interesting properties that can be harnessed. For more information about nanotechnology, check out Chapter 19.

Uncovering lanthanides and actinides

Buried deep inside the transition metals are two more groups with important, unique characteristics — lanthanides and actinides. They are unique because they use orbital shells that aren't important to the rest of the periodic table. The chemistry of these materials are not fully understood yet, because some are rare and hard to find, whereas others are radioactive and dangerous to work with. For more information about these elements, see Chapter 14.

Diving Deeper: Special Topics

In Part IV, you get the opportunity to explore some of the more specialized subfields of inorganic chemistry. Each chapter introduces you to how inorganic chemistry is used in a specific way, such as increasing reaction speed (catalysis), or capturing energy from the sunlight (in a chemical reaction called photosynthesis), and building smaller and smaller computer devices. In each chapter, we only brush the surface of these fascinating special topics. But you have enough of the working tools to further your own detailed study of these topics when you want.

Bonding with carbon: Organometallics

In Chapter 14, we introduce the field of *organometallic chemistry*. As the name suggests, it deals with the chemistry of carbon-containing (or organic) molecules called *ligands* that bond with metals to form organometallic compounds. Organometallic chemistry combines some aspects of organic chemistry with some aspects of metallic chemistry, and the results are compounds with some unique traits, such as:

✔ The effect of the ligands can be so significant that the colors can be bright blue, red, or green, depending on what ligands are used and where they are placed around the metal center. Atoms with the same metal center can have very bright and brilliant color changes with the addition of different ligands. Many of these compounds are used as pigments in paints.

✔ Most of the organometallic compounds are made with transition metals as the metal center. These metals can have differing magnetic properties depending on the oxidation states, which can be controlled by the placement and type of ligands that are used around the metal.

✔ Organometallic compounds are often used as catalysts. Because they can have very specific geometries, they can make very specific chemical reactions occur.

Speeding things up: Catalysts

Imagine how much more work you could get done if you found a short cut that's faster and has greater precision in producing results. In chemistry this is possible thanks to catalysis. *Catalysis* is the chemistry of making things happen faster, or making them happen with less required energy, or both. Catalysis is carried out by chemicals that are called *catalysts*. A catalyst makes light work out of heavy-duty chemistry. Catalysts are important because they allow for the quick and cheap production of strong and durable materials, such as plastics.

Inside and out: Bio-inorganic and environmental chemistry

You don't just find examples of inorganic chemistry in the laboratory or in industry; you can also find them inside yourself or around your environment. For instance, the oxygen you're inhaling right now is being transported around your body by an iron compound inside a large organometallic molecule called *hemoglobin*. In Chapter 17, we explain how and why this works. Other examples of bio-inorganic chemistry that are described in Chapter 18 include:

- ✔ **Photosynthesis:** The chemical reactions involved in photosynthesis transform sunlight energy and carbon dioxide molecules into sugar, water, and oxygen molecules.

- ✔ **Nitrogen fixation:** Some bacteria perform chemical reactions that capture atmospheric nitrogen and fix it so that it can be absorbed by organisms (usually plants) through a series of inorganic chemical reactions. The importance of this chemistry can't be over emphasized. Nitrogen is extremely important to living matter, and nature has developed efficient methods using enzymes in bacteria to work with nitrogen. Science has only recently created similar tools to do so, albeit much more crude than the way that nature does.

- ✔ **Enzymes:** Enzymes are proteins that act as catalysts for important functions within your body. Take for example, lactase — the enzyme that's used to help with the digestion of milk. Some people are lactose intolerant because they lack this enzyme, but they can overcome this by consuming a pill that contains lactase.

Solid-state chemistry

Solid-state chemistry is based on the study of atoms that combine to build solid structures, or crystals. In Chapter 18, you learn how solid-state chemists describe the shape of crystal structures and how this determines the size and shape of the unit cell, which is then used to characterize the many different forms that solid structures take. For example:

- ✔ **Simple crystal structures:** Simple crystal structures are composed of atoms that are positioned on the edges of the unit cell.

- ✔ **Binary crystal structures:** Binary crystal structures are made of two type of atoms in the crystal, such as NaCl (table salt), for example.

- ✔ **Complex crystal structures:** These are more involved than the other examples because they can have more than two different types of atoms present.

One of the most important advances in solid state chemistry is the development of silicon-based materials. The Silicon Valley is where the semiconductor industry was born; scientists worked very hard to learn how to purify silicon and arrange the silicon atoms in such a way that they can be used to make a computer chip. At the heart of every single computer, and most electronic devices, is silicon. Just look around you and imagine a world without silicon, it would be a very different place.

Nanotechnology

In the final chapter of Part IV, we tackle a very new and exciting field called nanotechnology. In this area of study, the size and shape of materials is often of paramount importance. At the size scale of living matter (bacteria are 20 nanometers (nm) in size, DNA is 1-2 nm wide) inorganic chemists can make exquisite materials with near-atomic precision. The advantage of nanotechnology is realized in many different applications; for example, it can be used to enhance catalytic processes, in biomedical applications, and to enhance the mechanical properties of bulk materials.

Nanotechnology is one of the most recent developments to arise from the sciences. It was developed only a couple of decades ago, but already the number of scientific publications and discoveries has been staggering. One of the unique features of this area is that it is important not just for the development of chemistry research, but also physics and biology, too. For this reason many new developments are occurring due to collaborations among researchers of physics, chemistry, and biology. Some of these include foldable electronics, anti-cancer treatments, ever smaller computers, and new methods of water filtration, to name just a few.

Nature has been working at the nanoscale for eons, only now can humankind begin to work at this scale, too. This final chapter gives a brief introduction to the major findings and applications of nanotechnology, but in no way gives full justice to the vast amount of work being carried out in this field. We hope that upon reading Chapter 19 you agree that the future is nano, and that inorganic chemistry plays a vital role in the continued development of this technology.

Listing 40 More

The last part of this book (Part V and the Part of Tens) gives you some non-technical information about inorganic chemistry. We start right at home by listing some of the common household products that involve inorganic chemical reactions, or inorganic compounds in Chapter 20. These household items may come in handy if you want to try out any of the ten experiments listed in Chapter 23.

Chapter 21 describes ten chemists (or teams of chemists) who have played an important role for inorganic chemistry and who were recognized for their achievements by receiving the Nobel Prize. Finally, Chapter 22 describes ten of the more useful and interesting techniques used in inorganic chemistry research.

Chapter 2

Following the Leader: Atomic Structure and Periodic Trends

*O*ne night, an electron, proton, and neutron came together to form an atom. They were so excited, they decided to go out for a fancy dinner to celebrate. When it came time to pay the tab, the three particles decided to have the bill split evenly amongst themselves. When the waiter returned, he handed the electron and the proton separate receipt books. Confused, the neutron asks: "Where's mine?" The waiter smirked and said, "For you, sir, *there's no charge.*"

This chapter explores how these three critical particles (neutrons, protons, and electrons) render the structures of the numerous atoms we interact with, as well as how repeating trends can be used to predict properties of unknown elements. Atoms are critically important to chemistry. Just as a tower made of Legos can be taken apart to its individual bricks, all of the molecules that make up *everything* around you can as well. These bricks can then be sorted by all the bricks that have the same properties: You could make a pile of small green pieces, small yellow pieces, and large red pieces, for example. Although a giant tub of Legos can make thousands of different designs, each of those designs stems from the same, limited number of unique pieces. Similarly, the Chemical Abstract Service (CAS) Registry, a list of known organic and inorganic substances, has over 64 million molecules and grows at a rate of 15,000 molecules a day all constructed from nature's limited number of building blocks — currently scientists have only discovered 118!

Up an' Atom: Reviewing Atomic Terminology

There are three *subatomic particles,* or particles smaller than an atom, that comprise the matter in the world around us. Everything we see, touch, smell, taste, and so on, is made of *atoms,* the basic building blocks of all matter. In turn, each of these atoms contain a combination of:

✔ **Neutrons:** Neutrally charged particles found in the nucleus of an atom.

✔ **Protons:** Positively charged particles, also found in the nucleus; it's important to note that the number of protons an atom possesses is the sole factor that distinguishes one element from another.

✔ **Electrons:** Negatively charged particles not found in the nucleus, but at the core of most all chemical reactions.

It's important to remember that an element is defined by the number of protons it has. For example, all carbon atoms, by definition, have six protons; however, many *isotopes,* or atoms with the exact same number of protons but different numbers of neutrons, of carbon exist. The most common three isotopes of carbon are carbon-12, carbon-13, and carbon-14. Each of these isotopes has six protons, yet a varying number of neutrons (six, seven, and eight, respectively). The sum of the protons and neutrons make up the *mass number.*

Figure 2-1 shows the *nuclear notation* (a way of writing elements that gives information about the nucleus of element) for the S_2^{2+} polyatomic ion with 16 neutrons on each sulfur atom, or nuclide.

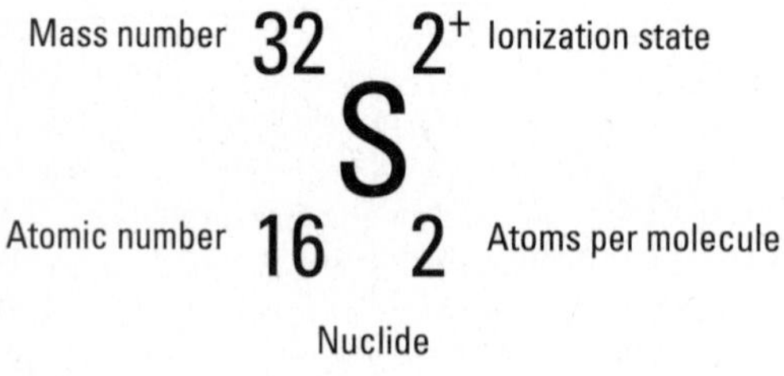

Figure 2-1:
Nuclear notation for S_2^{2+} polyatomic ion with 16 neutrons on each sulfur atom.

Mass number **32** **2**$^+$ Ionization state

$$S$$

Atomic number **16** **2** Atoms per molecule

Nuclide

Finding atoms

The story of the atom is a wonderful tale that has existed for centuries, before scientists could even prove that an atom, in fact, exists.

✔ **Around 450 B.C.:** Leucippus and his pupil Democritus develop the idea that all matter is made up of atoms, an ancient Greek word for indivisible, using the logic argument of everything can only be divided a finite number of times until it is too small to be further divided. At that point, everything is atoms (indivisible particles) and empty space. Moreover, the atoms must take on properties of their bulk materials: strong iron must have hooks that hold the iron atoms together, while water must be smooth to allow it to flow, for example. However, this theory took a backseat to Aristotle's classical five elements: earth, wind, water, air, and aether for many centuries.

✔ **1803 A.D.:** John Dalton notices that reactions occur in specific proportions based on the respective weights and determines the relative weights of six elements (H, O, N, C, S, and P). From this, Dalton created the first points of modern atomic theory:

Elements are made of atoms.

Atoms of any element are identical.

Atoms cannot be subdivided, created, or destroyed.

Atoms form compounds in whole-number ratios.

Chemical reactions are the rearrangement of atoms.

Although many of these postulates were later disproven, these underlie a major turning point in the chemist's view of the world,

earning Dalton the title as one of the fathers of modern chemistry.

✔ **1897 – 1904 A.D.:** Following the recently discovered negatively-charged electron (called "corpuscles by their discoverer J. J. Thompson), the plum pudding model of the atom was born. In this model, the atom remained a definite shape, like Democritus and Leucippus proposed, but with the majority of the space being positively charged with tiny specks of negatively-charged particles distributed throughout like plums throughout plum pudding (or like chocolate chips throughout a chocolate chip cookie).

✔ **1909 A.D.:** Rutherford and his team of scientists developed the idea of a positively-charged nucleus surrounded by mostly empty space after bombarding many metal foils, most famously his gold foil with alpha particles, making the curious observation that a tiny amount of the alpha particles bounced back. Rutherford famously described this to be as amazing as shooting a cannonball at a piece of tissue paper and having the cannonball bounce back. The only explanation was a highly concentrated nucleus of positive charge at the center of an atom.

✔ **1913 A.D.:** A Danish physicist named Niels Bohr envisaged electrons orbiting the nucleus in discrete orbits, where each orbit has a very specific energy level that could explain the different (but very specific) energies of light that were emitted from different elements. Essentially, this model mimics the planets of our solar system orbiting our sun, but relies on the force of attraction between the negative electrons and the positive nucleus instead of gravity like our solar system uses.

(continued)

(continued)

⊭ **1950s A.D.:** Quantum mechanics jazzes up all of these theories of the atom by showing that just as traditional wave-like light could be considered a particle, traditional particle-like electrons could be considered like a wave. This *wave-particle duality*, one of the central theories underlying quantum mechanics, allowed for the position of an electron to be calculated as the probability of finding an electron in a specific point around the nucleus.

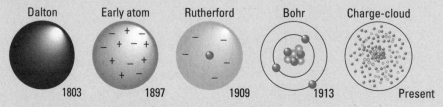

The various models of the atom through the ages.

Figure 2-1 has five key pieces of information about any given nuclide. Starting with the number in the lower left and going clockwise around the "S":

⊭ **Atomic number (lower left, 16):** This is the number of protons within the nuclide. As the atomic number tells you the element, which is also represented by the element's symbol at the center of the nuclear notation (in this case, S for sulfur), it's commonly not included. All atoms that have 16 protons are sulfur atoms.

⊭ **Mass number (upper-left, 32):** This represents the sum of the neutrons and protons present in the nuclide. To figure out the number of neutrons a nuclide has, subtract the atomic number (16) from the mass number (32).

⊭ **Ionization state (upper-right, 2+):** Remember that an *ion* is an atom that has gained or lost an electron, so it has a net electrical charge. If positive (+), the atom lost electrons; if negative (–), the atom gained electrons.

⊭ **Atoms per molecule (lower-right, 2):** This number simply tells you how many atoms are making up the molecule or polyatomic ion in question, just as the "2" in H_2O tells you there's two hydrogen atoms in each water molecule.

⊭ **Element's symbol (center, S):** Just like the atomic number (and also straight off the periodic table), this tells you what element your nuclide is.

All naturally occurring elements, from hydrogen to plutonium, are found in the rock, water, air, or earth. All the other elements found in the periodic table must be made using nuclear chemistry and are not found naturally, they are dealt with in Chapter 14.

Sizing up subatomic particles

One of the best analogies for visualizing the size of an atom is to take an orange and imagine that the orange were the size of the Earth. Now, as you are imagining that you are standing on a giant orange instead of the Earth, and the orange was the size of the earth then all the atoms that make up this orange would be about the size of a normal orange. Think of how many oranges it would take to make a sphere of oranges the size of the Earth.

Now that amazing size difference only gets you down to the size of the atom, but the particles within the atom are even smaller. Keeping with our orange analogy, if we make the orange a hydrogen atom (the smallest atom), the nucleus would be too small to see with the naked eye. In fact, if you blew up the orange to the size of the a major sports arena such as the infamous Astrodome in Houston, Texas, the nucleus would only be about the size of a pea sitting on the 50 yard line.

Figure 2-2 shows the relative size scale going even smaller than the nucleus; you quickly notice that the electron is to the nucleus what that nucleus is to the atom in size scale. Going beyond the protons, neutrons, and electrons usually is leaving the world of chemistry and entering nuclear physics, but it's interesting to note that protons and neutrons are made up of a triplet of particles known as *quarks*, and those particles are about the size of an electron. Having a good grasp on nuclear structure allows radioactive dating, cancer radiotherapy, nuclear weaponry, nuclear power, and countless other critically important applications of nuclear chemistry.

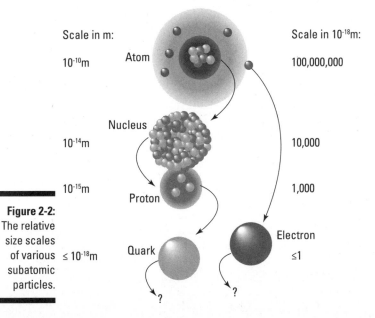

Scale in m:

10^{-10}m   Atom

10^{-14}m

10^{-15}m

Nucleus

Proton

Figure 2-2:
The relative
size scales
of various $\leq 10^{-18}$m   Quark
subatomic
particles.

Scale in 10^{-18}m:

100,000,000

10,000

1,000

Electron

≤ 1

?

?

Knowing the nucleus

The atomic nucleus was discovered in the 1900s by the work of several physicists, most notably of which was Ernest Rutherford and his team of scientists that included Hans Geiger and Ernest Marsden. Prior to the Geiger-Marsden experiment, sometimes just called the Gold Foil experiment, scientists knew that atoms had positive and negative charges and that the negative charges could be removed. The working model, known as the plum pudding model, treated the atom like plum pudding, with the tiny pieces of plum being the negative charges and the pudding being the positive charge. Having never eaten plum pudding, I like to imagine a spherical chocolate chip cookie where the cookie is positively charged, but the chocolate chips are little specks of negative charge distributed throughout the cookie.

The test involved bombarding various metal foils — most famously gold foil — with positively charged particles, known as alpha (α) particles resulting from newly discovered radioactive decay. When focusing the α particles into a stream and aiming them directly at a very thin film of gold foil, almost all of the α particles passed straight through the foil. This further supported the idea the plum pudding model: The small amount of positive charge the atom contained was dispersed over such a large area that it wouldn't have any meaningful interactions with the α particles. To better visualize this, imagine filling a sports arena with a material that when weighed was roughly the same as a bullet — you're going to need to fill that sports arena with something even less dense than packing noodles, because you are dispersing the mass over such a large area. The same is true for the plum pudding model, only using positive charge. Just as you would imagine shooting something lighter than Styrofoam all day long, every bullet would fly straight through barely being affected in the slightest, so would an α particle through the plum pudding model.

However, something amazing happened: Every so often an α particle could be detected bouncing back toward the source after striking the foil. Just as our Styrofoam analogy, Rutherford likened this rare event to shooting a cannon ball at a piece of tissue paper and having it bounce back at you. After lots of repetition to make sure the results weren't a fluke, a new model needed to be designed to account for how almost all the α particles could go straight through the material, yet a tiny amount would bounce back. This resulted in the Rutherford model, hinging on a very small nucleus of very dense, positive charge that occupied very little space relative to the atom (100,000 times smaller than the atom). Interestingly, that tiny nucleus would have to contain the majority of the mass of the atom so it would be very dense.

Going orbital

This new line of thinking about the atom (really dense nucleus with tiny particles moving around it) hit close to home (or rather, light years away from home) for many astronomers. If tiny planets in our solar system orbit around our massive sun, perhaps electrons could also orbit the nucleus of an atom. This so-called planetary model of an atom was a dominant theory for a while, until the strange quantum world was better understood.

Remember that one of the major points of quantum mechanics is that it's nondeterministic, or rather that the future can't be determined from knowledge of the present. In our world, the classical world, when I throw a ball at a window, I can calculate exactly where it will hit, how hard it will hit, the path it will take to get to the window, and so on. Quantum mechanics is more like flipping a quarter. Sure, the last nine flips may have all landed on heads, but that new flip still has only a 50 percent chance of landing on heads or tails, regardless of what it has previously done.

Because of this nondeterministic nature of the electrons, the electron can't follow a path like a planet would an orbit; we clearly can predict the movement of orbiting planets or else we wouldn't have solar eclipse parties (well, at least I wouldn't have solar eclipse parties . . .). However, the orbital theory worked because the electrons were found in discrete energy levels, suggesting that some sort of fixed distance travel: Imagine the nucleus as a big magnet that was attracting an electron (not too far from the truth). Now try and pull that electron away from the nucleus. If it was really close, you can imagine that it would be harder to pull that electron away than if it was further away to begin with. Well, if every time you tried to pull away that electron it took the same amount of manpower, it would be logical to assume that it was always the same distance away.

While the math underlying quantum mechanics supports the idea that an electron does not necessarily follow an orbit-looking path, there is a certain region in space where you are likely to find one if you went looking for it; this region is called *an orbital*. Here's another way to think about it: Although we can't know the exact position of an electron at all times, we can expect it to be somewhere inside our orbital a certain percentage of the time. Occasionally you encounter an orbital described as a cloud of electron density.

You didn't think I was going to let you get away without at least a few equations under your belt, did you? Okay, I won't make you solve it, but let's at least discuss what an orbital is all about. First, imagine an atom of hydrogen with just one lonely electron. While you can't form a predictive model that will "follow" an electron over time like you could plot a baseball after being hit by a bat, you can treat the electron as a wave (remember that electrons possess wave-particle duality). The easiest calculation (which by no means is easy) is to treat the wave as a *standing wave*, or a wave in a constant position. Perhaps you remember the big jump ropes in gym class that if you and your partner wiggled up and down just right, the jump rope formed a regular pattern — you may not have known it, but you made a standing wave.

Scientist Erwin Schrödinger developed a series of equations that were critical for developing *quantum mechanics*, a special type of physics that focuses on really small things (atoms or smaller) and how they can be both particle-like and wave-like in properties. Schrödinger focused on the wave-like properties of an electron to describe how these small things change over time. It turns out in the case of hydrogen, the electron orbital is spherical around the nucleus, as shown in Figure 2-3.

Figure 2-3:
The solution to the Schrödinger wave-function equation for a hydrogen atom gives the orbital shape of a sphere that circulates about the nucleus of the atom.

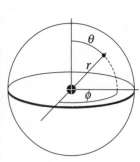

$$\hat{H}\,\psi(x) = E\psi(x)$$

$$\left[-\frac{h^2}{2m}\nabla^2 + V(x) \right]\psi(x) = E\psi(x)$$

$$\Psi(r,\,\theta,\,\phi) = R(r)Y(\theta,\,\phi)$$

To make matters simple, Figure 2-4 shows the results for all the various orbitals for the hydrogen atom. If you wanted to solve these orbitals for larger atoms, or especially molecules and compounds, you wouldn't be able to do them by hand. You would need large computing power to come to an acceptable answer for what the orbitals truly look like.

Following naming conventions

Modern quantum theory tells you that the various allowed states of existence of the electron in the hydrogen atom correspond to different standing wave patterns that are the electron orbitals.

- ✔ **Principal quantum number (n):** The shell number or row number.

- ✔ **Azimuthal quantum number (l):** The shape of the orbital, sometimes called the subshell.

- ✔ **Magnetic quantum number (m_l):** The specific subshell orbital. For example, there are three subshell orbitals for the p orbital: $p_x(m_l = -1)$, $p_y(m_l = 0)$, and $p_z(m_l = +1)$.

- ✔ **Spin quantum number (m_s):** This term describes axial spinning of an electron in the particle. The up and down arrow symbols are used to signify when the electron is spin-up ($m_s = +\frac{1}{2}$) or spin-down ($m_s = -\frac{1}{2}$).

Figure 2-4:
The electron orbitals around a nucleus. These graphical representations are the most likely places to find the electrons as they orbit the nucleus in their respective orbitals.

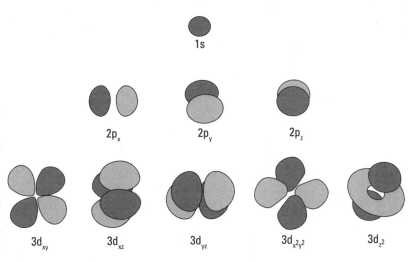

Each of the four quantum numbers (n, l, m_l, and m_s) has only a few allowed values:

- n can have only positive integer values ($n = 1,2,3,4,...$). The number 1 is the first row of the periodic table, 2 is the second row, and so on.

- l can have values ranging from zero to one smaller than the principal quantum number ($0,1,2, ... ,n–1$). There are four possibilities: s ($l = 0$), p ($l = 1$), d ($l = 2$), and f ($l = 3$).

- m_l can have values from $–l$ to l, so s can be 0, p can be –1, 0, or 1, d can be –2, –1, 0, 1, or 2, and f can be –3, –2, –1, 0, 1, 2, or 3. If you sum these up, s has one subshell orbital, p has three, d has five, and f has seven.

- m_s can be only –½ or ½.

The electrons that are nearest the nucleus are very tightly bound and so have a lower energy when compared to those of the electrons further from the nucleus as shown in Figure 2-5. For hydrogen, all of the subshell orbitals are degenerate (or have the same amount of energy) because hydrogen has only one electron. However, for all other elements, the presence of additional electrons leads to slight energy differences between the subshells; following the pattern $s < p < d < f$ with regards to the energy for those shells. For this reason, exceptions to the rule arise when there are cases having a smaller n; for example, notice how the 4s orbital is lower in energy than the 3d orbitals in Figure 2-5.

Figure 2-5:
Figure 2-5:
The energy
levels of the
electrons
around the
nucleus,
demonstrat-
ing how
the further
away the
orbital is
from the
nucleus,
the higher
energy it
will have.

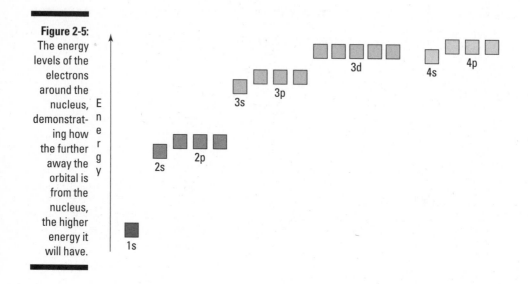

Fitting in electrons

The Pauli exclusion principle states that for each electron pair in a shared orbital, each electron must have an opposite spin state — only one spin-up $\left(m_s = +\frac{1}{2}\right)$ and one spin-down $\left(m_s = -\frac{1}{2}\right)$ electron for each orbital. Because of this rule, no two electrons in the same atom can have the same set of quantum numbers. If it weren't for the exclusion principle, the atoms of all elements would behave in the same way, and there would be no need for the science of chemistry (as we currently know it).

Distinguishing atomic number and mass number

The number of protons found within an atom's nucleus is known as its atomic number (Z). Knowing an atomic number is the same as knowing an element's name — they both identify the element. *Every* atom with an atomic number of 6 is carbon, and *every* carbon atom has six protons.

Just because all atoms of the same element have the same number of protons, however, doesn't mean they are all the same. One can add or subtract neutrons to an atom without changing the atom's atomic number, but the atom becomes heavier or lighter because neutrons have mass similar to protons. When two atoms are the same element but differ in mass, they are said to be *isotopic* (*iso* means the same, and *topic* means place, as in place on the periodic table). For two isotopes of the same element, the atomic number is the same; to distinguish these isotopes, the *mass number*, or sum of the *nucleons* (protons and neutrons), is used.

Quantum mechanics versus the Bohr Theory

The Danish physicist Niels Bohr won a Nobel Prize in Physics for developing a model of an atom where electrons moved around the nucleus in circular paths. The Bohr model of the electron orbiting the nucleus is a nice model to start with because it gave chemists and physicists alike something certain to picture in their minds. It helped to explain the relationship between elements and the colors that they make when excited. In the Bohr model the electrons orbit the nucleus sort of like the planets orbit the sun. But upon closer scrutiny it was discovered that the electron position and speed could not both be known with great certainty. This became known as the uncertainty principle, and it meant you could only guess at, or give a good approximation of, where you can expect to find an electron at any given time. So the electron position is based on the probability of where it's most likely to be found. This was a great discovery in quantum mechanics and led to many philosophical and technical debates. Today, it's considered the gold standard to understanding many processes in chemistry.

The most important result of quantum mechanics, the uncertainty principle, and electron probability is that we now have a picture of the electron orbitals that help scientists understand chemistry even better.

The Bohr model of the atom was correct in many ways but one, the electron was considered to be in one certain place at a time, (that it was a particle fixed in space). It was not until De Broglie declared in his PhD defense that in fact an electron can act as both a particle and a wave that some rethinking was required. Because if the electron acted like a wave, then it was not a fixed point in space. So the mechanics of motion of an electron would require solving a wave equation. Erwin Schrödinger, the same Schrödinger from the orbital section previously discussed in this chapter, came up with the wave equation for determining the position of the electron around the nucleus.

Identifying isotopes

There are three naturally occurring isotopes of oxygen: oxygen-16, oxygen-17, and oxygen-18. All three of these isotopes have an atomic number of 8, so these isotopes have eight, nine, and ten neutrons, respectively. Figure 2-6 displays these three isotopes of oxygen. Over 99.7 percent of all the oxygen are oxygen-16 due to the special stability that comes from having an equal number of protons and neutrons for smaller atoms. However as atoms get larger, they require more neutrons to remain stable. Therefore, small oxygen-16 (8 protons, 8 neutrons) and large lead-204 (82 protons, 122 neutrons) are both stable.

Isotopes all act the same chemically, because they have the same number of protons, which implies they have the same number of electrons, which determines chemical reactivity. Isotopes that are unstable undergo radioactive decay; they are called *radioisotopes*.

Figure 2-6:
There are three isotopes of oxygen, the most commonly found is oxygen-16 that is found with an abundance of 99.76 percent, this is a stable isotope.

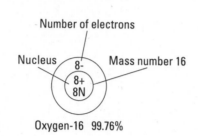

Oxygen-16 99.76%

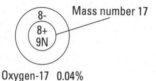

Oxygen-17 0.04%

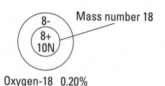

Oxygen-18 0.20%

Grouping Elements in the Periodic Table

The periodic table, shown in Figure 2-7, is probably one of the most important artifacts of science to date. In the simplicity and arrangement of the periodic table, all the intricacies of the material world can be understood and predicted. As the name suggests, the table displays all the elements according to their periodic arrangement.

There are two important terms to realize — *period* and *group*. The periodic table is like a matrix because there are columns and rows. The columns distinguish the groups. And the periodicity is shown in the rows. There are 18 groups, and 7 periods, and these can be seen along the top and along the side of the periodic table in Figure 2-7.

This is based on the filling of electrons in orbitals around the nucleus. The s orbital has two electrons (one pair), the p orbital holds six electrons (three pairs), the d orbital fits ten electrons (five pairs), and the f orbital can hold 14 electrons (seven pairs). A simple way to remember is one, three, five, and seven electron pairs, for the s, p, d, and f orbitals.

Because the outer orbitals and the number of electrons in that orbit determines the chemical differences, it is logical to order the periodic table according to "blocks" associated with a particular orbit. Hence there are s-block elements, p-block, and so on.

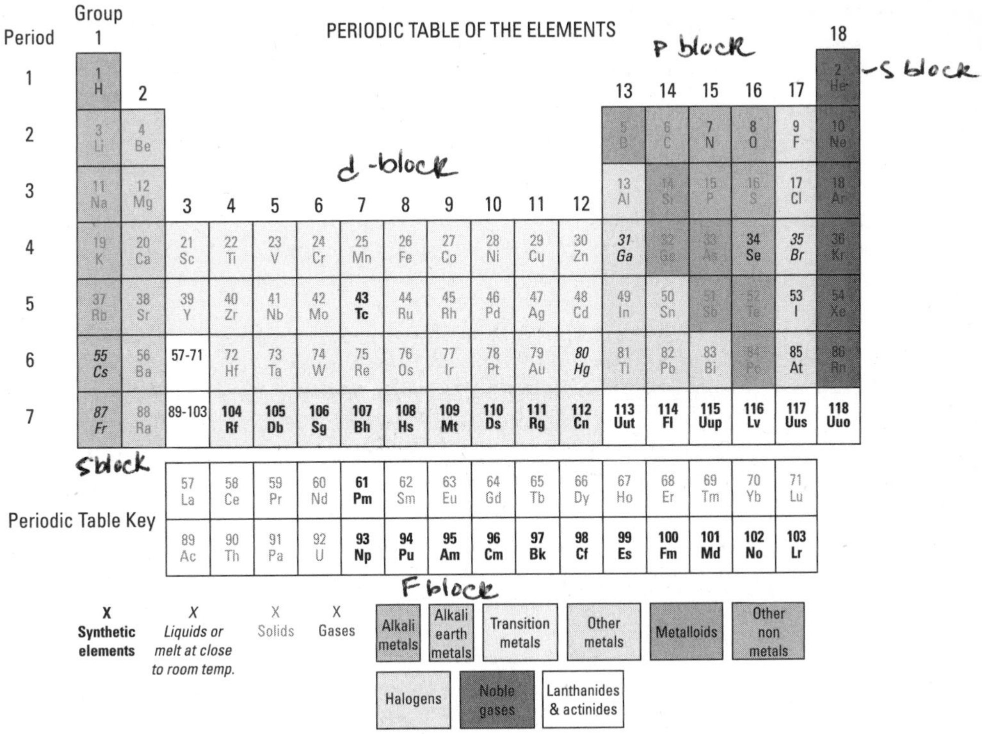

Figure 2-7: The periodic table.

Elements in the same block share some similarities, but the similarities are clearer when the elements are ordered according to groups.

Keeping up with periodic trends

The general (big picture) trends for the periodic table are shown in Figure 2-8. As you can see the atoms get bigger as you move down a column, and smaller as you move to the right across a row, or period.

The chemical properties of an element are largely decided by the electrons in the outer orbitals. This is largely associated with its valency. So when the elements are grouped according to the number of electrons in the outer orbitals, it's nearly enough to know the differences among each of the groups. Because the elements in the same group share similar properties, you don't have to learn about all of the elements (103 of them); instead you can start by learning the differences among the groups (only 18).

Valence is a term that signifies the power of the atoms to form chemical bond. It's Latin for "be strong."

Moving across the periodic table the number of the electrons in the outer orbital increases from 1 to 8. Elements in Group 1 have one outer electron, such as hydrogen. And elements in Group 2 have two outer electrons, such as beryllium; these elements are divalent. Basically, the s-block elements have the same valency number as the group number.

The p-block elements have a valency that is either the group number minus 10, or 18 minus the group number. This is equivalent to the number of electrons in the outer shell because subtracting by ten removes the ten d-block elements (that would be considered inner electrons).

Moving left to right across the periodic table in Figure 2-8, the atomic size decreases because of the additional nuclear charge due to addition of neutrons and protons in the nucleus. This makes the orbital electrons be more tightly bound, and the ionization energy (the amount of energy required to remove the outer most electron; see the section eyeing ionization later in the chapter for more information) increases because of that.

Moving left to right on the table the metallic character (properties that are associated with metals) also changes; it decreases from left to right (See Figure 2-8). Also, the oxides of these elements become less basic. For example, Na_2O is strongly basic, and Al_2O_3 is amphoteric (meaning it can react as an acid or a base), and SO_2 is an acidic oxide. In general, metal oxides are basic, and non-metallic oxides are acidic.

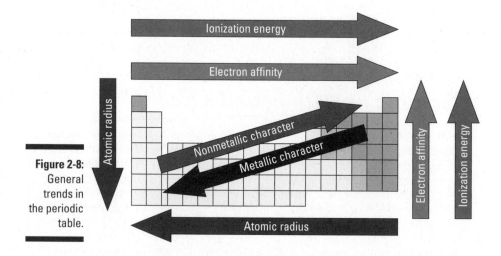

Figure 2-8:
General trends in the periodic table.

When you go down the periodic table from top to bottom in same group, all the outer electrons have the same valency, but the size increases. This leads to a decrease in ionization energies, and an increase in the metallic character.

This is most obvious with the Group 14 and 15 elements. Starting with carbon and nitrogen, you go down to metals such as lead and bismuth. Also, the metals become more basic going down in each group.

If you move down diagonally to the right from each element there are also some similarities to note. The similarities aren't that strong when compared with being in the same group, but for some pairs the similarity is quite strong. These are lithium to magnesium, beryllium to aluminum, boron to silicon. Further to this the line that distinguishes between metals and nonmetals run diagonally.

Measuring atomic size

Neutrons and protons are about 1,860 times heavier than electrons, yet the electrons cloud takes up most of the space of an atom. In fact, although the nucleus contains 99.95 percent of an atom's mass, it occupies only a trillionth of the volume.

If all the mass is so concentrated, it seems logical that scientists could just take an atom, put it on a tiny scale, and figure out what it weighs. However, when the periodic table was developed, scientists were not able to take atoms and weigh them out. Instead, chemist John Dalton did notice that in chemical reactions elements had to be mixed at very specific proportions. For example, to make H_2O you need the exact amount of hydrogen and oxygen mixed together. In this way you could work out, for example, that to make 100 percent H_2O you need a mass of hydrogen and a mass of oxygen, and you find that oxygen is 16 times heavier than hydrogen. This doesn't really tell you the weight yet, but it could give you a relative weight that can get you started. Because hydrogen is the lightest element, you can start comparing the weight of other elements to that of hydrogen, until finally you are able to measure the mass of one atom at a time.

You can also try calculate the size of an atom by using *Avogadro's number* (N_A). This number (6.022×10^{23}) represents the number of molecules found in a *mole* of a substance. While this mole sounds like the furry, little burrowing mammal, it's actually a quantity; just as the term "dozen" represents 12 and "a pair" represents 2, a mole represents 6.022×10^{23}. To demonstrate the amazing size of this amount, imagine that you have a mole of paper clips (or 6.022×10^{23} paper clips). If you linked those paper clips together, they would extend from the Sun to Pluto and back . . . over a billion times. If you have one mole of a substance, it has Avogadro's number of atoms. Divide the volume of that material by Avogadro's number and you find the value for each atom (but this is not the most accurate method for determining volume).

Rating the atomic radius

Measuring the size of an atom is not a straightforward answer either. It's like measuring a ball of fur. If you measure it as you find it, and then you squeeze it together, the ball of fur has different sizes. This is the same for atoms, because there are different ways to look at the size of the atom. And each of these ways use different techniques to get results. So depending on the technique, the answers vary.

You can estimate the size by calculation based on quantum mechanical principles, using solutions to the Schrödinger equation that helped develop the shapes of orbitals (see the section "Going orbital" earlier in this chapter). Figure 2-9 shows the results from solving the Schrödinger equation to solve for atomic diameter. You could also determine the size based on the *covalent radii*, the distance between to nuclei in a covalent bond; covalent radii can be determined using x-ray diffraction. Finally, you can determine the radius using the *van der Waal radii*, which is the distance between nonbonded atoms that are in contact with or touching other molecules.

Figure 2-9: Sizes of the atoms according to the atomic radii (first number), the covalent radii (second number), and the Van Der Waals radii (third number).

1.58 / 0.3 / 1.2																	0.98 / n.a.
4.10 / 1.52	2.80 / 1.12											2.34 / 0.88	1.82 / 0.77	1.50 / 0.70 / 1.5	1.30 / 0.66 / 1.40	1.14 / 0.64 / 1.35	1.02 / n.a. / 1.60
4.46 / 1.86	3.44 / 1.60											3.64 / 1.43	2.92 / 1.17	2.46 / 1.10 / 1.9	2.18 / 1.04 / 1.85	1.94 / 0.99 / 1.80	1.76 / n.a. / 1.92
5.54 / 2.31	4.46 / 1.97	4.18 / 1.60	4.00 / 1.46	3.84 / 1.31	3.70 / 1.25	3.58 / 1.29	3.44 / 1.26	3.34 / 1.25	3.24 / 1.24	3.14 / 1.28	3.06 / 1.33	3.62 / 1.22	3.04 / 1.22	2.66 / 1.21 / 2.0	2.44 / 1.17 / 2.00	2.24 / 1.14 / 1.95	2.06 / n.a. / 1.97
5.96 / 2.44	4.90 / 2.15	4.54 / 1.80	4.32 / 1.57	4.16 / 1.41	4.02 / 1.36	3.90 / 1.3	3.78 / 1.33	3.66 / 1.34	3.58 / 1.38	3.50 / 1.44	3.42 / 1.49	4.00 / 1.62	3.44 / 1.4	3.06 / 1.41 / 2.2	2.84 / 1.37 / 2.20	2.64 / 1.33 / 2.15	2.48 / n.a. / 2.17
6.68 / 2.62	5.56 / 2.17	5.48 / 1.88	4.32 / 1.57	4.18 / 1.43	4.04 / 1.37	3.94 / 1.37	3.84 / 1.34	3.74 / 1.35	3.66 / 1.38	3.58 / 1.44	3.52 / 1.52	4.16 / 1.71	3.62 / 1.75	3.26 / 1.46	3.06 / 1.4	2.86 / 1.4	2.68 / n.a.
2.7 / 2.20	2.20	2.2															

Eyeing ionization energy

↪ The *ionization energy* is the energy necessary to remove an electron from an atom: The higher the ionization energy, the stronger the electrons are bound to an atom. As you move left to right across a period of the periodic table, the first ionization energy generally increases due to the decreasing radius of the atoms. Think of pulling on a really large magnet near your refrigerator — it's easy to move if you are a few feet away, but takes a lot of strength to rip away when only inches apart. The same is true for the electrons — opposite charges attract (negative electrons to the positive protons of the nucleus). Going down a group of the periodic table, the atoms get larger, so the ionization energies decrease from top to bottom.

Each electron in an atom has an ionization energy, so the energy required to remove the first electron is known as the *first ionization energy*, the next electron is known as the *second ionization energy*, and so on. After removing one electron, the electron-electron repulsion is smaller in the electron cloud surrounding the nucleus, so the electron cloud contracts a little making the ion a little smaller. As stated before, the smaller the atom, the higher the ionization energy.

Following this trend, each subsequent ionization energy is higher in energy. If you monitor these energies, not only do they increase at every new ionization energy, but eventually you see a very large jump; for example, the first four ionization energies for aluminum are: 578 kJ/mol, 1,817 kJ/mol, 2,745 kJ/mol, and 11,577 kJ/mol, respectively. The massive jump in energy occurs between the third and fourth ionization energies. Aluminum, having only three valence electrons, is electronically like a noble gas after removing three electrons, so removing the fourth electron (the fourth ionization energy) requires a large amount of energy because you are destroying the *very* stable noble gas configuration.

kJ/mol is a unit of energy per an amount of material. The kJ stands for kilo Joule, where kilo is a unit of thousands, and Joule is a unit of energy. Another way of reading this is to say it has 587 kilojoules of energy per each mole of the molecule. Mole and mol refer to the same property. It is more convenient to write mol when writing equations instead of mole, and so it has been adopted as the accepted symbol for the mole.

There are a few other anomalies in the general trend of ionization energies increasing left to right, and those are seen going from Group 2 to Group 13 (or from two valence electrons to three) and from Group 15 to Group 16. To understand these jumps, it's important to remember our orbitals: There's one *s* orbital capable of having two electrons, and three *p* orbitals each capable of having two electrons. In Group 2, both valence electrons fill the *s* orbital (remember, the filled *s* orbital is a stable noble gas configuration for hydrogen). Because of the stability of the filled *s* orbital, there's a slight decrease in ionization energy going from Group 2 to Group 13. The second anomaly is going from Group 15 to Group 16, when the *p* electron count goes from three to four. As you add electrons, there's a slight amount of energy required to have two electrons paired together in the same orbital, an energy known as *pairing energy*. When you add the first three *p* electrons to the three *p* orbitals, each electron can claim its own *p* orbital without having to pair up. The minute you add the fourth electron, it must pair up with another electron, slightly weakening the electron attraction to the nucleus (and in turn, slightly lowering the ionization energy).

Examining electron affinities

Opposite to ionization energy, *electron affinity* is the amount of energy an atom or ion releases when an electron is added. When an atom takes on an electron, general energy is released, which would be a negative change in energy (ΔE), so electron affinities are usually negative. To complicate matters even further, for many atoms it is easier to determine the ionization energy (energy to remove an electron) for the atom with an extra electron. In other words, if electron affinity is: $X + e^- \rightarrow X^-$, then you could do the reverse $X^- - e^- \rightarrow X$. By doing the opposite, the change in energy would be positive, not negative. In general, nonmetals have higher magnitudes for electron affinities than metals.

Noting electronegativity

If you take the average of an atom's ionization energy and electron affinity, you have a measure for the likelihood of an atom attracting an electron, known as *electronegativity*. The higher the electronegativity, the more an atom would like an electron. This mathematical method of calculating electronegativities is known as Mulliken electronegativity, named after its Nobel Prize–winning discoverer, Robert Mulliken. Another Nobel Prize–winning scientist, Linus Pauling, came up with a similar method that was calculated using the relative strengths of bonds in molecules; for example, if you knew the bond energy for hydrogen (H-H), chlorine (Cl-Cl), and hydrogen chloride (H-Cl), then you could determine which element was more likely to attract an electron relative to one another. If you continue this calculation through the entire periodic table, you would have the Pauling electronegativity scale (which ranges from 0.7 for the least electronegative francium to 4.0 for the most electronegative fluorine).

Regardless of which scale you use, (there are numerous others not even mentioned, though Pauling and Mulliken are the most commonly encountered) the general trend is the same: Electronegativity increases going from the bottom to the top or from the left to the right on the periodic table. The general trend is shown in Figure 2-8.

Chapter 3

The United States of Oxidation

*O*ne of the corniest jokes you will find in chemistry goes like this:

> *An atom walks into a bar, and the bartender says: "You look upset! Is everything okay?"*
>
> *"Yeah, I'm a little sad," responds the atom. "I've lost an electron."*
>
> *"Are you sure?" asks the bartender.*
>
> *"Yeah, I'm positive" responds the atom.*

This joke hinges on the fact that when an atom loses an electron, it becomes a positively charged ion (also known as a cation). Jokes aside, this chapter is all about electronic bookkeeping. In chemistry, electrons underlie almost all reactions, whether they are being excited, forming bonds, and so on. Oxidation states are simply a way to keep track of where all those electrons are going.

Entering the Oxidation-Reduction Zone

Oxidation and reduction are processes that describe when an atom transfers an electron. If the atom gains an electron, it's reduced; if the atom donates an electron, it's oxidized.

There are two primary mnemonics (memory aides) to help remember which is which:

✔ OiL-RiG: Oxidation is Loss, Reduction is Gain.

✔ LEO the lion says GER: Loss of an Electron is Oxidation, Gain of an Electron is Reduction.

If mnemonics aren't your thing, you can also simply remember that the charge of the atom that gets reduced is decreased. For example, copper(II) chloride (also known as cupric chloride or $CuCl_2$) has copper in a +2 oxidation state (Cu^{2+}). Forming copper (I) chloride (also known as cuprous chloride or $CuCl$) involves changing the copper to Cu^{1+}. In short, +2 was reduced to +1; therefore, the copper was reduced. Note that this isn't the historic background to the term *reduction*. Historically, oxidation used to refer to an element that reacted with oxygen, and reduction referred to a metal that was produced from its mineral ore.

Oxidation occurs when an element combines with an element that is more electronegative, reduction with an element less electronegative. Complete oxidation-reduction reactions, known as redox reactions (*redox* is a portmanteau of reduction and oxidation just like *smog* is a portmanteau of smoke and fog), include both processes to ensure everything remains balanced, or rather if something loses an electron something else must gain an electron. Redox reactions are chemically interesting because they describe how a variety of things work, ranging from batteries, rusting bike tires, to nerve cells.

Redox reactions are the principle way that nearly all the elements are purified after they have been extracted as minerals.

✔ *Oxidation* refers to gaining oxygen, losing hydrogen, or losing electrons. It's an increase in the oxidation number. It's also an increase in the valence number of the atom.

✔ *Reduction* refers to losing oxygen, gaining hydrogen, or gaining one or more electrons. It is a decrease in the oxidation number. It's also a decrease in the valence number of the atom.

✔ The *oxidation state* describes the degree of oxidation of an atom. Though bonds can never be 100 percent ionic, an oxidation state represents the charge that would exist if all bonds on an atom were purely ionic (given, not shared). Oxidation states range from negative values to positive values and are not required to be integers (though they usually are). They are written as a symbol followed by a number, such as −1, +3, or +7.

✔ A *valence number* or the *valency* describes how many bonds an atom can form. Valency is not seen as commonly as oxidation states, because oxidation numbers tend to offer the same information as a valence number as well as a better description of the atom's electronic state.

✔ The *oxidation number* is used in coordination compounds and represents the charge an atom would have if all ligands and electron pairs were removed from the central atom. Oxidation numbers are usually identical to oxidation states, and the terms are often used interchangeably. The key difference is that a true oxidation number is represented by a Roman numeral. For example, iron with an oxidation number of +2 could be written as iron(II) or Fe^{II}.

Following oxidation state rules

It's often difficult to keep track of where all the electrons come from and end up, so a system of rules based on oxidation states exists to help keep everything straight.

> **Rule #1:** The oxidation state of a free or pure element is zero (0). This includes monoatomic (Fe, Co, Ni, and so on), diatomic (Cl_2, F_2, Br_2, and so on) and even polyatomic (S_8) molecules of the same element.

> **Rule #2:** Monoatomic ions have an oxidation state equal to their charge. For example, chloride anion (Cl^{-1}) has an OS of –1 and calcium cation (Ca^{2+}) has an oxidation state of +2.

> **Rule #3:** With rare exceptions, hydrogen has an oxidation state of 1, and oxygen has an oxidation state of –2. Exceptions to this rule are:

> **a.** Hydrogen has an oxidation state of –1 in hydrides (LiH, BH_3, AlH_3, and so on).

> **b.** Oxygen has an oxidation state of –1 in peroxides (R-O-O-R, such as H_2O_2) and an oxidation state of –½ in superoxides (O_2^-, such as NaO_2).

> **Rule #4:** The sum of the oxidation states of all atoms in a molecule must be zero if neutral, or equal to the charge of the ion if there is a net charge.

The oxidation state is the charge left on the atom after all the other atoms of the compound have been removed. Some elements can have multiple oxidation states, such as Tl (+3 and +1), Sn (+4 and +2), and P (–3, +5, and +3).

Regardless of whether a bond is covalent or ionic, the oxidation state can be calculated. The oxidation state of S in sulfuric acid (H_2SO_4) can be calculated through the oxidation rules above. Rule #3 makes each of the four oxygen atoms have a oxidation state of –2, and each hydrogen a +1. Overall it's known that the total molecule has no charge, so all the oxidation states must add up to 0. With these facts, you can easily determine the oxidation state (OS) of S:

$$OS_H + OS + {}_sOS = 0 \text{ o}$$
$$2*(-1) + OS + 4(-2) = 0 \text{ s}$$
$$OS_S = +6$$

So by following the rules, you can deduce that the oxidation state of S is (+6).

In the case of ionic compounds such as potassium permanganate ($KMnO_4$), the potassium is ionized into K^+ cations that counter the negative charge of the MnO_4^- anion. You can therefore solve for the oxidation state of Mn two separate ways:

Use the entire molecule, like you did for H_2SO_4:

$$OS_K + OS_{Mn} + OS_O = 0$$
$$1*(+1) + OS_{Mn} + 4(-2) = 0$$
$$OS_{Mn} = +7$$

Use just the anion because you know the charge on the anion:

$$OS_{Mn} + OS_O = -1$$
$$OS_{Mn} + 4(-2) = -1$$
$$OS_{Mn} = +7$$

Notice that both methods result in the correct oxidation state for manganese (+7).

Transition metals can have several oxidation states, traditionally increasing or decreasing by units of 1 or 2.

Examples include Fe^{3+} and Fe^{2+}, Cu^{2+} and Cu^+, and Pt^{2+} and Pt^{4+}. This is specific to the d-block elements, because certain d-electrons can partake in bonding.

The main group elements show a periodic pattern of oxidation states. For the first 25 elements, there is a stepwise increase in the oxidation state for the elements as you follow them going across the periodic table, left to right, as shown in Figure 3-1. The maximum oxidation state of an element is equal to the number of electrons in the outer orbital for these first 25 elements.

For example, aluminum is $[Ne]3s^23p^1$ (three outer electrons) with an oxidation state of +3, but bromine is $[Ar]4s^23d^{10}4p^5$ (seven outer electrons) and has an oxidation state of +7.

Many nonmetals and semimetals have more than one oxidation state, such as nitrogen with oxidation states that range from –3 to +5. Also, the common oxidation states of nonmetals tend to decrease in units of two; for example see chlorine +7, +5, +3, +1, –1. This pattern of two is due to a pair of electrons that remain as a pair and yet do not form any bonds (the inert pair effect).

For the s-block of the periodic table (Groups 1 and 2), the oxidation state is always the same as the group number, except for that pesky hydrogen –1 exception.

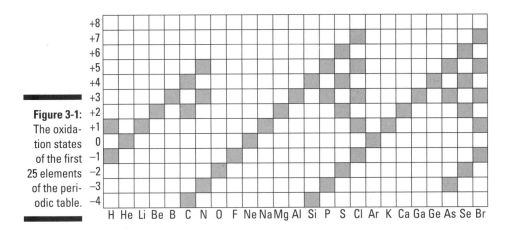

Figure 3-1: The oxidation states of the first 25 elements of the periodic table.

Scouting reduction potentials

The *reduction potential* (E, measured in volts) of an element measures the tendency of an element to be reduced (or gain electrons). This potential can tell you what types of reactions are possible for that element, what other elements it can bond with, and what types of bonds will likely be formed. To better grasp a reduction potential, you need to first understand a half-reaction. Essentially, a half-reaction will show how much energy it takes to reduce an element.

To start, consider an electrochemical cell, or a type of electrical battery pictured in Figure 3-2. Notice that this battery has two terminals immersed in liquid solutions:

✔ Anode: the electrode where current flows *into* the solution, marked as the (–)-terminal.

✔ Cathode: the electrode where current flows *out of* the solution, marked as the (+)-terminal.

Conventional electrical current flows in the *opposite* direction as the electron flow. Because of this, when current is flowing into a solution at the anode, electrons are flowing out of the solution. Remember, if electrons are being pulled away from the metal atoms of the anode, they are losing electrons (oxidation). Therefore, oxidation occurs at the anode.

Figure 3-2:
The reduction potential of a material can be determined using an electrolytic cell. A salt bridge of known concentration is used to induce the electron transfer.

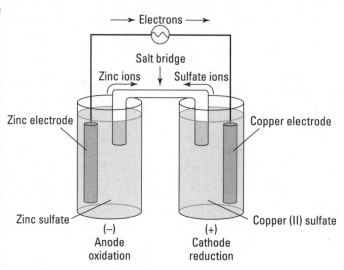

Some helpful mnemonics for remembering anodes and cathodes are:

✔ **RED CAT, OX AN: RED**uction occurs at the **CAT**hode; **OX**idation occurs at the **AN**ode

✔ **R**oman **CAT**holic **AN**d Orthodo**X**

✔ **ACID: A**node **C**urrent **I**nto **D**evice

✔ **CCD: C**athode **C**urrent **D**eparts

The battery generates electrical energy from the redox reactions that occur within the cell. It generally consists of two half-cells, each containing different metals dipped into ionic solutions with similar cations (for example, a copper cathode in copper(II) sulfate and a zinc anode in zinc sulfate). Then, these two half-cells are connected by a salt bridge. A salt bridge simply connects the two half-cells so the charged ions can flow from one half-cell to the other. These can be glass tubes filled with inert electrolytes (usually just table salt and water) or simply a piece of paper soaked in the electrolyte. The salt bridge helps keep the charge balanced, completing the circuit without letting the solutions mix together.

The half reaction with the more negative reduction potential occurs at the anode as oxidation, while the half reaction with a more positive reduction potential occurs at the cathode, as reduction.

To calculate the reduction potential of a cell, you can use the following formula:

$$E^{\circ}_{cell} = E^{\circ}_{cathode} - E^{\circ}_{anode}$$

The symbol ° denotes that the reaction is performed at standard conditions: 25 °C, 1 M concentration, pressure of 1 atm, and metals in their pure state. Everything is set relative to a standard hydrogen electrode that is assigned a reduction potential value of 0.00 V.

Consider the reaction of Cu^{2+} with H_2, which occurs spontaneously, generating electron flow.

$$Cu^{2+}(aq) + H_2(g) + H_2O \rightarrow Cu(s) + 2\,H_3O^+(aq)$$

If you allow this reaction to run, you could simply measure the voltage using a voltmeter, and determine that the cell has a voltage of 0.34 V.

Because the potential of the hydrogen is defined at 0.00 V, the half-reaction for $Cu^{2+}/Cu = +0.34$ V, relative to hydrogen.

Another way to write this is:

$$Cu^{2+}(aq) + 2\,e^- \rightleftharpoons Cu(s)\quad E° = +0.34\ V$$

The reduction potential cannot always be calculated for a given element, but it can be compared with that of hydrogen. If a substance is easier to reduce than hydrogen, then the E° is positive. But if it is harder to reduce than hydrogen, the E° is negative.

Using a spontaneous reaction as an example, you can determine from the electric current between two electrodes how many volts are produced from the transfer of electrons. When measured, you can determine how many volts are required to reduce or oxidize an element.

Platinum (Pt) is often used as a passive electrode because it is very unreactive and will not form compounds with the elements in solution. Pt is unreactive due to having a very high ionization energy, yet is a rare element and quite expensive. Gold and graphite can also be used.

Examining the cell in Figure 3-2, it is possible to determine the voltage of the battery given the half-cell reactions that occur. If you acquire a table of known half-cell reactions, you will find the following listed:

$$Cu^{2+}(aq) + 2e^- \rightleftharpoons Cu^0(s)\quad E° = +0.34\ V$$
$$Zn^{2+}(aq) + 2e^- \rightleftharpoons Zn^0(s)\quad E° = -0.76\ V$$

Remember that the table only gives you the reduction potentials, yet the galvanic cell has both a reduction reaction *and* an oxidation reaction. For a galvanic cell, you want a positive voltage, so the cathode is the metal with the more positive reduction potential, in this case copper. Reduction occurs at the cathode ("red cat"), so the half reaction is written correctly. Oxidation, on the other hand, occurs at the anode ("ox an"), so the Zn half-reaction is written in the wrong direction. As written, it shows that Zn^{2+} is being reduced to Zn^0, when you would want the opposite to occur as you want the copper to be reduced. By writing the equation backwards, it's important to note that the sign of the potential changes:

$$Zn^0(s) \rightleftharpoons Zn^{2+}(aq) + 2\ e^- - E° = +0.76\ V$$

Combining the reduction reaction with the oxidation reaction yields the overall reaction of Cu^{2+} ions and zinc metal making copper metal and Zn^{2+} ions with a total potential of 1.10 V:

$$Cu^{2+}(aq) + 2\ e^- \rightleftharpoons Cu^0(s) + 0.34$$
$$\underline{Zn^0(s) \rightleftharpoons Zn^{2+}(aq) + 2\ e^- + 0.76}$$
$$Cu^{2+}(aq) + Zn^0(s) \rightleftharpoons Cu^0(s) + Zn^{2+}(aq) + 1.10$$

Notice that you are forming copper metal and removing zinc metal; over time, you would expect that the copper cathode will get heavier (due to the Cu^{2+} ions in solution depositing as copper metal) and the zinc anode will get lighter due to the zinc metal being oxidized and dissolving into the solution.

To prevent corrosion, such as rusting, many engineers will add zinc to metal structures to intentionally act as *sacrificial anode*. When your bicycle starts to rust, the Fe metal of your bike frame is getting oxidized to iron cations while the oxygen in the air and water around your bike frame gets reduced to OH^- anions. When these iron cations and hydroxide anions meet, rust is produced. If an engineer installs zinc into the system, zinc will be oxidized instead of the iron, which can prevent the formation of rust. Because the anode gets consumed, it is called a sacrificial anode.

Walking through a Redox Reaction

In a redox reaction, the total charge is always balanced. In the simplest of cases, one atom gains electrons, while another loses electrons. As with the galvanic cell, the reaction can be divided into two half reactions, one of oxidation and one of reduction. These half reactions can make balancing redox equations easier when many elements are involved.

Consider the following equation where the numbers above the chemical symbols represent the oxidation states of the elements.

$$\overset{1,-1}{HCl} + \overset{4,-2}{MnO_2} = \overset{2,-1}{MnCl_2} + \overset{1,-2}{H_2O} + \overset{0}{Cl_2}$$
$$+ 2e^-$$
$$-1e^-$$

Hydrogen and oxygen have the same oxidation state on both sides of the equation (1 and –2, respectively), yet both manganese and chlorine change (for example, the oxidation state of Mn decreases by 2, so it gained two electrons; Cl ions loses one electron). However, this isn't an even exchange of electrons as written. For this to balance out, there need to be two atoms of Cl (so two electrons are lost to equal the two electrons gained by Mn). Thus the equation looks like:

$$2\,HCl + MnO_2 \rightarrow MnCl_2 + H_2O + Cl_2$$

At this point, the electron exchange is balanced; however, there are four Cl atoms on the right and only two on the left. Finally, changing the coefficient of HCl to 4 and H_2O to 2 balances out the elements.

$$4\,HCl + MnO_2 = MnCl_2 + 2\,H_2O + Cl_2$$

Now the equation is balanced and complete.

In the following example, however, the redox equation only lists the ions important to the reaction (spectator ions, or ions not involved in the actual reaction, are not listed).

$$\overset{2}{Fe^{++}} + \overset{7}{MnO_4^-} + H^+ = \overset{2}{Mn^{+++}} + \overset{3}{Fe} + H_2O$$
$$\underset{+5e^-}{\underline{\hspace{2cm}}}$$
$$\underset{-1e^-}{\underline{\hspace{3cm}}}$$

Five Fe^{2+} ions will have to react with each reactive permanganate in order to balance the gain and loss of electrons. To do this, place a coefficient of 5 in front of each iron ion:

$$5\,Fe^{2+} + MnO_4^- + H^+ = Mn^{2+} + 5\,Fe^{3+} + H_2O$$

Again, the electron exchange is balanced, but not the elements (the oxygen and hydrogen atoms are not equal on each side of the equation). Balancing the elements results in this equation:

$$5\,Fe^{2+} + MnO_4^- + 8\,H^+ = Mn^{2+} + 5\,Fe^{3+} + 4\,H_2O$$

The equation seems to be balanced, but to balance ionic reactions you need to ensure the net charge on each side of the equation is the same. In this case the net charge +17 appears on either side; therefore, the equation is balanced.

Isolating Elements

The majority of the elements on earth are found naturally in combination with other elements in materials such as minerals and ores. It is necessary to extract these elements from their ores and minerals to use them. Most of the elements are found within compounds in a charged oxidation state (most often positive) such as Ti^{4+}, Zn^{2+} and Cl^{-1} in TiO_2, ZnS, and $NaCl$, respectively. Purifying these elements requires redox chemistry, and careful selection of the appropriate choice of reductants and oxidizers is important.

Carbon is a cheap and convenient reducing agent for metal oxides, so it used most often. However, it does not work for all the metals. Some metals are so electropositive (like aluminum), they require industrial electrolysis because the oxide form is so stable. If you think about it, when was the last time you saw a coke can rust? In fact, it took science so long to figure out how to extract aluminum (a common element in the earth's crust), that it was incredibly valuable in metallic form. Napoleon was rumored to reserve his aluminum flatware for only his best guests, and underlies the reason why aluminum was chosen for the tip to the Washington Monument in Washington D.C. Other elements, such as titanium, form stable carbides with carbon, so they require other methods of extraction, as well.

Figure 3-3 lists the common ways of extracting elements.

Mechanically separating elements

Some of the elements can be found in their native state because they are largely unreactive. Unfortunately, this list is short of metals that exist in their native state in large amounts; it includes gold, silver, and copper (the Group 11 elements) and the platinum group metals (ruthenium, rhodium, palladium, osmium, iridium, and platinum).

Gold is found as nuggets and is considerably more dense compared to the rocks they are found nearby. For this reason, panning is one of the oldest and simplest ways for mining gold. Simply throw a handful of gravel in a pan and gently shake the pan in water. The less dense rocks will leave the pan as you shake it, while the more dense nuggets of gold will remain in the bottom of the pan. Occasionally, silver and copper can be found in nugget form also. Because these three metals can be found easily in a native form, they were the basis of early metal coins, and hence are commonly referred to as *coinage metals*.

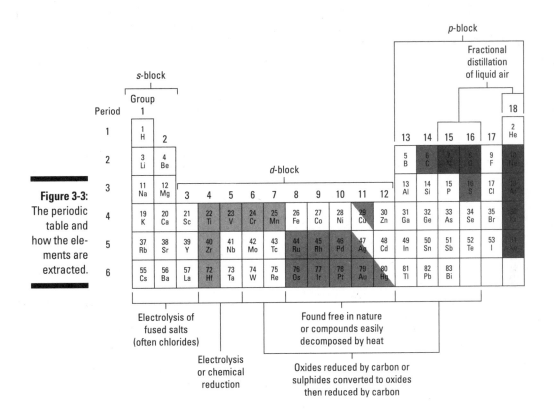

Figure 3-3:
The periodic table and how the elements are extracted.

Non-metals that are found naturally include S, Se, Te, C, Si, as well as the N_2 and O_2 in our atmosphere. The gases can be separated using fractional distillation of the liquid air, a method that uses different boiling points to separate them. Helium is also obtained from some natural gas deposits, a result of radioactive decay from elements such as uranium, yet is not naturally found in our atmosphere as it is too small and light and easily leaves our atmosphere when released. Think twice before you let go of that helium balloon, for after you let go, you will never see that He again! Diamonds, a form of carbon, are found naturally, and lead to big business for the mining of diamonds. Diamonds are also obtained by mechanical separation of rock. Graphite, another form of carbon, is also mined and mechanically separated. Deposits of sulfur are also found deep underground in Louisiana. Unfortunately, the towns that mine them are often reminiscent of the odor associated with sulfur (rotten eggs). Other elements, such as selenium and tellurium, can occasionally be found in these deposits.

Using thermal decomposition

Using high temperatures, some elements decompose into their elemental form, such as hydrides. Another example is sodium azide (NaN_3) that decomposes to sodium and pure nitrogen. This is an interesting compound to work with because it is an explosive, so usually only small amounts are made at any given time. This is commonly used for generating N_2 for laboratory work.

Nickel carbonyl ($Ni(CO)_4$) is a gaseous compound and can be used to create metallic nickel when heated through a process known as the Mond process.

Iodide compounds are not very stable, so heating them can result in an easy recapture of the metal elements in the compound. For example, the van Arkel-de-Boer process (or crystal bar process) is used to purify zirconium. This is done by passing the heated gas (ZrI_4) over a white hot tungsten filament, and over time the metals form on the tungsten.

Most metal oxides are stable up to 1000 °C, but metals below hydrogen in the electrochemical series (a series listing materials in order of their chemical reactivity with the most reactive on top, such as Ag_2O and HgO decompose with greater ease.) Iron, however, is by far the most widely separated element by means of thermal decomposition.

Displacing one element with another

Based on the electrochemical series, it is possible to displace any element in a compound with another element that is higher in the series. Displacement is often — but not always — used in conjunction with electrolysis. To make it economical, it is best to sacrifice a cheap element for a more expensive or rare element. For example, copper is replaced with iron, cadmium is replaced with zinc, and bromine is replaced with chlorine.

Heating things up: High-temperature chemical reactions

The most widely used method for preparing solid materials of pure form is by reacting them at high temperatures (in the correct molar proportions) over an extended period of time. Generally, temperatures of 2000 °C are achieved using furnaces or ovens and specialized glassware.

Carbon is the most common reductant in use, which is used in the form of coke — a cheap and readily available material. The reduction by carbon is used to generate iron, zinc, phosphorus, magnesium, and lead.

Additionally, another metal can also be used to reduce elements, as commonly seen with alumina (Al_2O_3). The oxidation of aluminum gives off a large amount of energy, in what is known as the thermite reaction. This reaction is used to fuel the flame of underwater welding machines, for example. Some of the elements purified in this way include manganese, boron, and chromium.

Relying on electrolytic reduction

An important observation of the redox reaction is that there is a transfer of electrons from the reducing agent to the oxidizing agent. Remember, a flow of electrons is the basis for the electrical current. Essentially, the strongest reducing agent is the electron itself.

In electrolysis, a positive electrical potential between an anode and a cathode separates elements. Examples include the use of NaCl (table salt) where molten sodium (reduced sodium cation) forms on the cathode, and Cl_2 (oxidized chloride anion) forms at the anode.

In cases where electricity is expensive, materials maybe shipped great distances for cheaper electricity sources. For example, large amounts of aluminum were reduced using cheap electricity in Norway, due to Norway's very large fjords and cascades that produce cheap power in large hydroelectric plants.

Chapter 4

Gone Fission: Nuclear Chemistry

In This Chapter

▶ Focusing on electrons

▶ Changing elements through radioactive decay

▶ Powering the world with nuclear energy

Remember those childhood birthday parties where there were *way* too many kids hopped up on birthday cake and ice cream for the chaperones to keep under control? It was only a matter of time before someone escaped the chaperones' watch and colored on the walls with crayons or crawled on top of the refrigerator to find the cookie jar. In many ways the nucleus, or the center of an atom where you find the neutrons and protons, is similar to these birthday parties — get too many protons or neutrons in the wrong ratio and one is bound (excuse my pun) to get away. This chapter explores radioactive decay, which is the process that atomic nuclei undergo to lose energy, similar to how the party naturally loses steam after the first kid leaves or more parents show up.

Noting Nuclear Properties

The majority of chemistry focuses on an atom's electrons and how they interact with one another (bonding, orbitals, and so on). All these interactions, however, don't hold a candle to the powerful interactions that occur within the atom's nucleus. If you need proof, go stand outside in the sunlight for a few minutes. The heat that you feel is the result of our solar system's massive nuclear reactor, a.k.a. the Sun! Thankfully, that reaction is situated millions of miles away; otherwise we wouldn't exist. In fact, nuclear reactions are so powerful they can even transform elements into different elements — an Alchemist's dream (too bad it costs more to synthesize gold in using nuclear chemistry than the gold is worth!).

The *atomic number (Z)* is equal to the number of protons (positively charged nucleons) in an atom, whereas the *mass number (A)* is equal to the sum of the protons (Z) and the neutrons (neutral nucleons, N) found in an atom's nucleus. For those who are partial to equations: $A = Z + N$.

Using the force

While the majority of inorganic chemistry functions off the same basic principle that leads to a fun blind date — opposites attract, you know — there are other forces at play in the nucleus of an atom. The concept of electromagnetic forces helps to explain the attraction of negatively charged electrons to the positively charged protons of the nucleus, and also why particles of the same charge repel one another (when you add an electron to an atom, the newly formed anion, or negatively charged ion, is larger than when it started because the electrons want to be as far apart as possible). This all begs the question: If opposites attract and like charges repel, why do all the positively charged protons cluster together in the center of a nucleus?

The *strong force* is another fundamental force, just like electromagnetism, that exists in nature and describes how particles interact with one another. As the name implies, this force is strong — about 100 times stronger than electromagnetic forces. This strength helps explain why the protons are willing to hang out in the nucleus despite having the same charge. Imagine at our birthday party that all the kids smell really bad (okay, not too much of a stretch . . .) — the kids might not want to hang out together due to the smell. Now add a gallon of ice cream to the middle of the room, and give each of the kids a spoon. Even though one force says they don't want to all be together (they're stinky), the stronger force (ice cream) can overcome it. Another principle associated with these fundamental forces is one of *range*. Some forces, like electromagnetism and gravity, have an infinite range; for example, even though the Sun is millions of miles away, the Earth is still attracted to it. The force does get weaker the further away you get (specifically, $1/r^2$ where r is the distance away the particles are), but it'll always be felt. The strong force, on the other hand, has a finite range, and beyond that range the force is not felt at all. For example, imagine that the ice cream was sitting on top of a big table that the children can't see unless they crawl up into a chair. If the kids do get close enough to crawl up into the chair and see the ice cream, no smell in the world will be strong enough to scare them away; however, if they are running around on the floor and never see the ice cream, it has no attractive force. At that point, stinky kids are to be avoided.

When the strong force is felt (or the ice cream is seen), the force does not change in strength (the kids want the ice cream just as much if it's a foot away or if it's ten feet away — after it's felt, it is all or nothing). Knowing this, it's important to be aware of the approximate range of this force to understand why atoms will undergo radioactive decay, because if the range is very large, all nucleons would *always* be content/stable. For example, no kid would ever leave the table after seeing the ice cream; however, what happens when the table is too small for all the kids to fit around? Only those kids who are lucky enough to be in a seat around the table are attracted to the ice cream, and the poor kids who lost our game of musical chairs have no attraction to the ice cream (because they can't see it) and are repulsed by the smell of the kids at the table (because that smell truly is unavoidable).

The empirical strikes back

These competing forces helped form the *semi-empirical mass formula (SEMF)* shown in the following equation:

$$E_B = a_V A - a_S A^{2/3} - a_C \frac{Z(Z-1)}{A^{1/3}} - a_A \frac{(A-2Z)^2}{A} + \delta(A,Z)$$

where A is the total number of nucleons (protons and neutrons), Z the number of protons, and N the number of neutrons. And where the nuclear binding energy (EB) is a measurement of the energy that's required to separate the neutrons and protons found in the nucleus into individual parts.

 This isn't the only time the concept of binding energy is used in chemistry. You also hear of *atomic binding energy,* which is the total amount of energy required to remove *all* the electrons from an atom, as well as an *electron binding energy,* which is the total amount of energy required to move a specific electron from an atom.

Notice that the SEMF equation shown earlier in this section has five distinct terms:

- ✔ **Volume term** ($a_V A$)**:** If you have more neutrons and protons, you have more strong force present. a_v is simply a constant; as *A* increases, so does amount of binding energy due to the volume term.

- ✔ **Surface term** ($a_S A^{2/3}$)**:** This term is nearly identical to the volume term, except that it accounts for protons and neutrons on the surface of the nucleus. Imagine you're at the supermarket and you're staring at a huge pyramid of oranges in the produce aisle. All the oranges in the center touch roughly the same number of oranges as one another (this is similar to the volume term); the oranges on the surface of the pyramid clearly touch less oranges because there are now portions of the orange that don't touch other oranges. In a similar fashion, this surface term accounts for the fact that some of the neutrons and protons interact with less particles due to being positioned on the surface of the nucleus. Also, similar to the volume term, a_s is simply a constant, so as *A* increases, so does the amount of binding energy due to the surface term.

✔ **Coulomb term** $\left(-a_C \dfrac{Z(Z-1)}{A^{1/3}}\right)$: The first thing you probably noticed about this term is that it's a negative term, meaning it accounts for a phenomena that *weakens* the attraction of the protons and neutrons. This goes back to our original assumptions: like charges repel each other! Whereas the strong force is stronger than the charge repulsions, it's important that you account for the repulsion when doing your energy bookkeeping. Notice that the number of protons has a much larger effect than the number of particles.

✔ **Asymmetry term** $\left(-a_A \dfrac{(A-2Z)^2}{A}\right)$: This term is a bit tricky at first; however, after you notice that $[N-Z] = (A-2Z)^2$ because the $A = Z + N$. Wait, what…? All that's really important about this term is that it derives from $[N-Z]$, or simply the difference between the number of protons and neutrons. Remember when you add electrons one by one to an atom and you fill an energy level, you have to add the subsequent electrons to a higher energy level because electrons can't occupy the same state (known as the *Pauli exclusion principle*); the same applies for protons and neutrons. Knowing this, if you have 1 atom with 1 proton and 19 neutrons, that 19th neutron is in a higher energy state than the 10th neutron if I had 10 protons and 10 neutrons; however, both atoms have a total of 20 protons and neutrons. So in short, this term accounts for the differences in energy when the number of protons doesn't equal the number of neutrons. Also, it's important to note that the term is all over A, which means that as the nucleus gets larger and larger, this term becomes more insignificant.

✔ **Pair Interaction Term** ($\delta(A,Z)$): This final term is similar to the asymmetry term. In short, the nuclear spin of the particles has two possibilities: spin-up or spin-down. Just like in the asymmetry term, there's a net benefit when all the spins are paired (the same number of spin-ups and spin-downs). The only way to have a perfect pairing of spin-up and spin-down is to have an even number of particles, so when Z and N are both even, you have more energy than when either Z or N is odd, which has more energy than when both Z and N are odd.

It's important to keep track of your signs when attempting to solve the SEMF equation. Remember that anything that strengthens the attractions between the particles is positive (or increases the binding energy) and anything that weakens the attractions is negative. The volume term and surface term both involve the strong force that strengthens the attractions, so they are positive terms. The coulomb term involves the repulsion between the positively charged protons that weakens the attractions, so it is a negative term. The asymmetry and pair interaction terms, on the other hand, are a little more tricky. They hinge off of a basic assumption: Your atom wants to have equal numbers of protons and neutrons, but most likely has an odd number of either Z or N. This translates into the asymmetry term is always zero unless Z doesn't equal N, in which case the asymmetry weakens the attractions causing the binding energy to go down. Similarly, the pair interaction term assumes a starting point of one term (either

Z or N) is odd. If they're both even, the attraction is even higher (as all spin-ups and spin-downs are balanced), so you add binding energy. If they're both odd, the attraction is even lower, so you subtract binding energy.

The SEMF is called *semi-empirical* because it's based partially on theory and partially on *empirical*, or information gained by means of observation or experimentation, data. There are different values you can find for the various constants used when calculating binding energies using the SEMF, but most solutions show that most stable isotopes (most strongly bound) is ^{63}Cu, very close to the empirical data for one of the most stable isotopes: ^{62}Ni. Figure 4-1 shows a graph displaying the binding energy per nucleon of various isotopes.

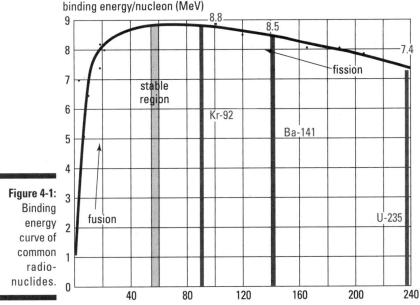

Figure 4-1: Binding energy curve of common radio-nuclides.

All elements with atomic numbers greater than 83 are radioactive, but knowing whether or not a nuclide is stable below an atomic number of 83 can be tricky. Figure 4-2 demonstrates how the stability of a nucleus depends on the ratio of the number of neutrons to the number of the protons. Notice how this figure starts out fairly linear, but then requires more neutrons than protons to remain stable. This occurs because when the nuclei is small, the asymmetry term wants $Z = N$; however, as the nucleus gets larger, the strong force can't affect nucleons far away (remember, after the kids are so far away from the ice cream and they can no longer see it, they simply forget about it). At this point, the electric repulsion takes over, so more neutrons are required to keep the positively charged protons away from one another. This belt of stability can be used to determine what kind of radiation decay likely occurs (for example, nuclei with too many protons want to emit a positron, or get rid of positive charge, to try and get closer to the band of stability).

Nuclei that have 2, 8, 20, 50, or 82 protons or neutrons (known as *magic numbers* — no, I'm serious) are stable. If the nuclei has a magic number of protons *and* a magic number of neutrons, it is said to be *doubly magic* (I couldn't make that up if I tried).

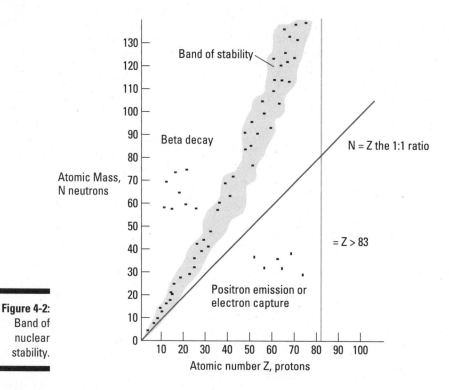

Figure 4-2:
Band of nuclear stability.

Documenting Atomic Decay: Radioactivity

Now that you know why all the nucleons stick together, here comes the fun part: learning why they all fall apart. Ironically, the *Godzilla, Teenage Mutant Ninja Turtles,* and *Spider Man* themes that might cover the children's gift bags are all supposedly due to the cartoon characters' encounter with radioactivity (and though it's not explicitly stated, I'm fairly certain *The Care Bears* came from radioactivity as well). Authors commonly use radioactive decay as a cause for super powers in stories because unstable nuclei, also known as *radionuclides*, release their extra energy in the form of *ionizing radiation*, or radiation that is powerful enough to rip the electrons off a surrounding atom or molecule. This radiation is similar to light rays, yet invisible to the naked eye. If this ionizing radiation comes into contact with one's DNA, the DNA molecules can actually undergo mutations; in our world, this usually leads to cancer — in comic books, this leads to *The Incredible Hulk.*

A French scientist, Professor Henri Becquerel, discovered that when he accidently placed a container of pitchblende (a mineral containing uranium) on photographic film, the film developed as though it were sitting in the light. Later, he placed a key under the container, and an image of the key appeared on the film when it was developed. Strangely, it appeared as though something similar to light was coming from the pitchblende. In 1903, Becquerel won half of the Nobel Prize in physics, "in recognition of the extraordinary services he has rendered by his discovery of spontaneous radioactivity" (while the other half was shared by Pierre and Marie Curie "in recognition of the extraordinary services they have rendered by their joint researches on the radiation phenomena discovered by Professor Henri Becquerel"). Later it was discovered that radioactive materials emitted energy by three possible routes. These were named after the first three letters of the Greek alphabet: alpha (α), beta (β), and gamma (γ) rays. Figure 4-3 shows the penetrating depth of these three forms of radioactivity.

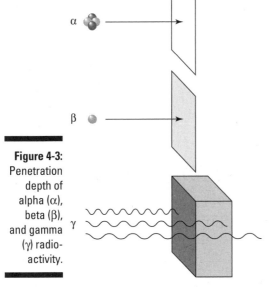

Figure 4-3:
Penetration depth of alpha (α), beta (β), and gamma (γ) radioactivity.

As you can see, γ radiation is the most penetrating, and often the most destructive (γ radiation is to blame for the increased rates of leukemia following the bombing of Hiroshima and Nagasaki, for example); however, this doesn't mean that α radiation is off the hook. If you recall the ex-KGB agent that was poisoned via radioactive sushi, he was killed by a very powerful α emitter. Sure, you can protect yourself from the α particles with a piece of tissue paper, but not from the inside! The penetration depth of a β particle is about 5 mm in solid aluminum. Materials such as lead, with high atomic numbers, are very good at stopping beta particles.

Alpha radiation

Alpha decay is denoted by the symbol α (Greek: alpha), and the radionuclide essentially just emits an atom of helium with no electrons $^4_2\text{He}^{2+}$; see Chapter 2 for isotope symbolism.

Remember that a proton or neutron is about 1,800 times larger than an electron, so an α particle with two neutrons and two protons is absolutely huge. Additionally, as this helium nuclei has two protons and no electrons, it's positively charged. Radionuclides that are rich in protons desire to undergo α decay to decrease their positive charge. For example, uranium-238 with 92 protons undergo a decay to become thorium-234 with 90 protons. Notice how the mass number went down by four (four nucleons are in an α particle), yet only two protons were lost (so the atomic number only decreases by two). As far as radioactive decay goes, alpha radiation is pretty slow; it can only travel a measly 10 percent the speed of light.

Another example of α decay occurs with radium-222 when it decays to radon-218 and an emitted α particle.

$$^{222}_{88}\text{Ra} \rightarrow {}^{281}_{86}\text{Rn} + {}^4_2\text{He}$$

Using a nuclear equation, you can calculate that the mass number on the product side (222 for radium) equals the sum of the mass numbers on the products side (218 for radon and 4 for helium). Likewise, the atomic numbers add up.

When alpha decay occurs, it stabilizes the nuclei that lie on the right of the band of stability (Figure 4-2). Upon α particle emission, the nucleus moves diagonally (left and down) toward the belt of stability as *both* the number of protons *and* neutrons are decreased by two.

Beta radiation

Beta decay is denoted by the symbol β, (Greek: beta) and it's symbolized as $^{\ 0}_{-1}\text{e}$. In many ways, β radiation is the opposite to α radiation.

β radiation is usually just the emission of an electron, and so it's negatively charged whereas α particles are positively charged; however, there's one form of β decay known as $β^+$ decay, or sometimes positron decay, where a radionuclide emits a positively charged electron known as a *positron*. A positron simply has the mass of an electron with the charge of a proton. Also, because electrons (and positrons) are so much lighter, they can travel much faster — almost as fast as light, though not quite. Regardless, it's certainly faster than α radiation. Finally, a β particle (when formed) actually leaves behind a nucleus with an extra proton, because it's created when a neutron decomposes. This neutron decomposition results in a photon (particle of light) and an electron (or β particle). So now the atomic number increases by 1 (instead of decreasing by 2 like α decay).

For example, thorium-234 (90 protons) undergo β decay to render proactinium-234 (91 protons) and an electron.

$$\ce{^{234}_{90}Th} \rightarrow \ce{^{234}_{91}Pa} + \ce{^{0}_{-1}e}$$

In this equation, you can see that β decay has the net effect of changing a neutron into a proton. It's important to remember that this β particle isn't an electron from the atom, but rather formed from the nuclei's neutron when it decayed. A neutron is approximately 1.675×10^{-27} kg, whereas a proton is only 1.673×10^{-27} kg, which helps explains how both an electron and a neutron can appear from β decay.

$$\ce{^{1}_{0}n} \rightarrow \ce{^{1}_{1}p} + \ce{^{0}_{-1}e}$$

Another form of β decay includes positron emission, where the nucleus expels positively charged positrons. For example, positron emission from a sodium nucleus looks like this:

$$\ce{^{22}_{11}Na} \rightarrow \ce{^{0}_{1}e} + \ce{^{22}_{10}Ne}$$

These particles don't last long before finding a negatively charged electron. As soon as these lovebirds meet, they run into one another at full blast and completely annihilate. All of the mass found in both the positron and electron is converted into pure energy (in the form of gamma rays) according to Einstein's famous $E=mc^2$. Doctors currently use low doses of positron-emitting radionuclides to image certain diseases using an instrument known as Positron Emission Tomography (PET). Similar to other tomographic imaging systems (such as computed tomography, or CT scan), this instrument monitors the gamma rays that come through the patient following their dose of radioactivity to triangulate the position of the radionuclide in the body. The formation of gamma rays follows this equation:

$$\ce{^{1}_{0}e} + \ce{^{0}_{-1}e} \rightarrow 2\,\ce{^{0}_{0}\gamma}$$

Another form of β decay includes electron capture, when an inner-orbital electron is captured by the nucleus and a gamma ray is produced:

$$\ce{^{201}_{80}Hg} + \ce{^{0}_{-1}e} \rightarrow \ce{^{201}_{79}Au} + \ce{^{0}_{0}\gamma}$$

In this example, the mercury-201 radionuclide gobbled up a surrounding electron from its external electron cloud and converted a proton into a neutron. Just how you saw that a neutron can decay into a proton and an electron, the reverse is also possible:

$$_1^1 p + \, _{-1}^0 e \; \rightarrow \; _0^1 n$$

Gamma radiation

Gamma radiation is denoted by γ (Greek: gamma) and is simply a high-energy photon of light. Occasionally, you encounter a gamma ray written as $_0^0 \gamma$, indicating that $Z = 0$ and $A = 0$.

γ radiation is unique from both α and β radiation because it's not a particle. It's more or less a packet of light called a photon, and so it is referred to as a γ ray just like a ray of sunshine. This little bundle of light has extremely high energy and can be created in the most energetic and violent of circumstances. For example, astronomers look for γ radiation in the cosmos because it's often a sign of a supernova explosion.

γ radiation doesn't cause the nucleus to decay into a different element. Instead of the nucleus being converted to another element like in α and β radiation, now the entire nucleus just sheds energy. Gamma radiation has no mass because there are no protons left, and it has no electricity. γ radiation travels at the speed of light — because it *is* light!

An electron that comes from a nucleus is called a β particle, and an x-ray that comes from the nucleus is called a γ-ray. Physically, these are identical; however, the names are different to denote where they originated from.

The half-life principle

Radioactive materials emit energy, such as a uranium rod used in a nuclear power plant. α and β radiation creates both thorium and neptunium from the uranium in the rods. Just as if you drive your car until the tank is empty, eventually even a nuclear power plant can run out of fuel. After a period of time, there simply will be none of the uranium left, and the rod will be composed of thorium and neptunium instead.

So, how long would this take for a nucleus of uranium to decay? Well, the answer is . . . no one knows. Nuclear decay is spontaneous, so it's impossible to determine when any one atom will decay. What can be determined, however, is the rate of decay for many nuclei. To do this, simply time how long it takes for half of the uranium rod to become something else. This time is called the radionuclides' half-life ($t_{1/2}$), and every radionuclide has its own unique $t_{1/2}$ that can stem from nanoseconds to billions of years. A half-life can be determined for any radionuclide given its decay rate, lambda (or λ):

$$t_{1/2} = \frac{\ln 2}{\lambda} = \frac{0.693}{\lambda}$$

The $t_{1/2}$ of uranium-239 is about 23 minutes, yet the $t_{1/2}$ of carbon-14 is about 5,700 years. In each $t_{1/2}$ cycle, half of the material transmutates while the other half remains the same. After the second cycle, only a quarter (a half of a half) of the original material is still the same. After the third half-life, only an eighth, then a sixteenth, and so on. Figure 4-4 demonstrates this decay for a radionuclide with a 29-year half-life.

Radioactive decay follows a first order rate law, because the concentration of the material is decreased by a factor of 2.

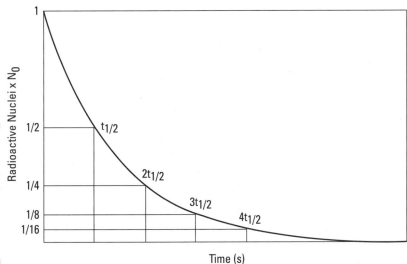

Figure 4-4:
The decay rate of a strontium radionuclide with a half-life of 28.8 years.

Blind (radiocarbon) dating

Radiocarbon dating is based on the conversion of radioactive carbon-14 to nitrogen-14 through β decay:

$$^{14}_{6}\text{C} \rightarrow ^{14}_{7}\text{N} + ^{0}_{-1}\text{e}$$

The amount of carbon-14 is constant in our atmosphere because it's continuously being produced in the atmosphere by the transmutation of nitrogen atoms with high energy neutrons:

$$^{14}_{7}\text{N} + ^{1}_{0}\text{n} \rightarrow ^{14}_{6}\text{C} + ^{1}_{1}\text{H}$$

You, your dog, and all the kids at our birthday party are made up of carbon, and you're all constantly replenishing that carbon through breathing and eating. The majority of carbon atoms are not radioactive, but every now and then a carbon nucleus has a few extra neutrons than normal (roughly one in every trillion carbon atoms). Carbon-14 consists of eight neutrons and six protons, whereas the majority of carbon is carbon-12 with six neutrons and six protons. Because both of these radionuclides are carbon (they have the same number of protons) they are chemically identical in reactions, so your body can't tell the difference.

When things die, they stop eating and breathing (though I like to think Uncle Frank is still snacking on a chicken wing somehow . . .); as such, the carbon-14 is no longer replenished in the system. Over time, the carbon-14 decays into carbon-12 (over a *lot* of time with a $t_{\frac{1}{2}}$ equal to 5,700 years). As the carbon-14 content decreases while the carbon-12 content increases, the ratio of carbon-14 to carbon-12 changes over time:

$$\frac{^{14}_{6}C}{^{12}_{6}C} > \frac{^{14}_{6}C}{^{12}_{6}C}$$

$$\quad\text{today} \qquad\qquad \text{fossils}$$

Using carbon-14 dating, scientists have dated some fossils back hundreds of thousands of years. There are, however, some drawbacks to carbon-14 dating. Primarily, the assumption is throughout history the amount ratio of carbon-12 and carbon-14 has been the same. Some evidence suggests that this ratio has slightly shifted over the ages, which could prevent truly reliable results if not taken into consideration. Another drawback to this technique is that in the lab a scientist has to take a pretty large size sample (0.5 g – 2 g); then it must be burned. For a tiny bone, that could require burning the entire bone just to know how old it is.

Radioisotopes

Isotopes of an element, such as chlorine, have the same number of protons, but a different number of neutrons. For example, there are two isotopes of chlorine, chlorine-35 and chlorine-37. These are both stable isotopes and have the exact same chemical reactivity because the number of protons (and hence electrons) are the same in each atom. When the ratio of protons to neutrons is beyond a certain limit, as shown in Figure 4-2, the energetic nucleus is said to be unstable; at this point, the isotope is radioactive and is called either a radioisotope or radionuclide.

Because radionuclides have the same chemical properties as their stable counterparts, scientists can sneak radioactivity into mixtures that are chemi-cally understood. For example, iodine is known to go to the thyroid and can be used for the treatment of goiter (a swollen thyroid usually due to an iodine deficiency — this is why you eat iodized salt). Doctors have used this fact to sneak iodine-131 to the thyroids of people with hyperthyroidism (overactive thyroid) because it behaves like stable iodine-127, yet applies therapeutic radiation.

In 1911, Ernest Rutherford was the first person to change one element into another. He did this by bombarding nitrogen with high-energy α particles, creating an isotope of oxygen and a proton.

$$^{14}_{7}\text{N} + {}^{4}_{2}\text{He} \rightarrow {}^{17}_{8}\text{O} + {}^{1}_{1}\text{p}$$

Since this first transmutation, several elements have been created in a similar fashion. Perhaps the most notable scientist in this field was Glenn T. Seaborg, who sought to create as many elements as possible and is now immortalized on the periodic table with element 106 — seaborgium.

Catalyzing a Nuclear Reaction

A nuclear reaction is a spontaneous process. In nature, you can find naturally occurring radionuclides that were created in the birth or death of a star light-years away from planet Earth. Some radioisotopes occur so rarely that huge effort has to go into their purification. Separating a stable isotope from chemically identical radioisotopes for use in a nuclear reactor is an energy-intensive process. Lots of material, such as pitchblende for example, has to be mined in order to purify radioisotopes of uranium for use in a nuclear power plant. Once again, this is because they are naturally occurring and occur so rarely.

There exists a need to overcome natural limits of radioactivity purification methods so that you can have a nuclear reaction on demand. To do this, scientists use high-energy collisions and literally smash atoms together as hard as current technology allows, hence the term *atom smashers*. It is quite an achievement to aim such small particles at each other and at such high speeds. These experiments take huge amounts of energy, but the energy that you can get out of them can be very large also, as evidenced by the blast of an atomic bomb or from the heat energy that comes from the sun.

When atoms are smashed together and undergo a transmutation, they can either split into smaller atoms or combine together to form larger atoms. When atoms split up and become smaller, the process is called *fission*, and the atoms literally just split into smaller pieces. Conversely, when the atoms come together to form a larger atom, the process is called *fusion* because the final product is simply two atoms fused together.

Fission and fusion are different from other radioactive decay modes because they won't happen spontaneously. For this reason, they need some kind of catalyst or special conditions to make them happen. Examples include the heat from the extremely high temperatures found in the center of the Sun or the force of using high-velocity particles to smash atoms together.

Atoms *want* to be stable. Knowing that, you can guess whether an atom will want to undergo fission or fusion by looking at Figure 4-1. The peak of that curve (around nickel and iron) are the most stable isotopes in existence (as they have the highest binding energy holding the nucleons together). Knowing that, elements to the left of the peak want to undergo fusion in the hopes of getting larger and getting closer to becoming iron. This is what's occurring right now in our sun — as it continues to burn up as fuel, it's forming larger and larger atoms. Conversely, elements to the right of the peak, such as large atoms of plutonium or uranium, want to get smaller to increase their binding energy, so they are likely to undergo fission. Just as with the kid's birthday ice cream, everything comes down to energy in the end.

Fission

Fission occurs when the atomic nucleus becomes overcrowded with nucleons, and the nucleus is split up into smaller parts, as shown in Figure 4-5. This can happen when it's struck by an accelerated particle, such as a neutron. In most cases, the resulting nucleus (now with an extra neutron and lots of energy) is unstable, so it undergoes a spontaneous radioactive decay. In doing so, the neutron is ejected, and the nucleus splits into two or more pieces; in many cases, even more neutrons are emitted in the process. The emitted neutrons can then collide with nearby nuclei and cause fission to occur again. This process naturally upscales (one neutron becomes many, and each of those neutrons becomes many, and so on) into a special reaction known as a *chain reaction*. The chain reaction occurs continuously as long as each collision results in at least one more neutron being emitted. If less than one neutron is emitted, then the reaction, now called a subcritical reaction, eventually stops. If exactly one neutron is emitted, the reaction becomes critical and the reaction rate is self-sustained. Perhaps most frighteningly, if each neutron results in multiple neutrons being emitted, the chain reaction is said to be supercritical and the reaction accelerates, and it could run out of control (just like the kids to the ice cream).

Three German chemists in the 1930s — Drs. Otto Hahn, Lise Meitner, and Fritz Strassman, just in case you were wondering — detected barium after bombarding uranium with neutrons from a neutron gun known as a *neutron howitzer*. The barium residue suggested that the team had actually fissioned the uranium nuclei. Later, this neutron bombardment was utilized in the first nuclear reactor capable of creating large amounts of energy from very small amounts of matter. This is the working principle behind many nuclear power stations: In a nutshell, the heat from the nuclear reaction is used to create steam, which is then used to move turbines to generate electricity.

Nuclear Fission

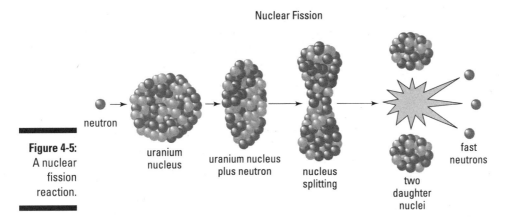

neutron

Figure 4-5:
A nuclear
fission
reaction.

uranium
nucleus

uranium nucleus
plus neutron

nucleus
splitting

two
daughter
nuclei

fast
neutrons

Fusion

Conversely, a fusion reaction occurs when two high-energy particles collide and are fused together (thus the creative name, fusion). To do this, small nuclides are used, such as deuterium (D or ^{2}H) and tritium (T or ^{3}H), which are both heavy isotopes of hydrogen (deuterium has one neutron and tritium has two neutrons). By fusing together two D atoms (which resembles an addition reaction), an atom of helium is created. As more atoms fuse together, heavier atoms can also be created. Figure 4-6 shows two nucleons of deuterium undergoing a fusion reaction. Note how there is a large release of energy as helium-3 is formed. This reaction also releases a neutron that can collide with another nucleus to keep fusion going in a chain reaction.

Interestingly, the mass of the helium atom is less than the mass of the two deuterium atoms, which at first glance seems to violate the conservation of mass. The difference in mass one would expect and the mass one actually sees is known as the *mass difference*. The mass difference is not actually missing, but rather converted to energy, about 100,000,000 kilocalories per kg of deuterium. When compared to burning gasoline, this reaction releases 10,000,000 times more energy. Unfortunately, fusion is an incredibly difficult process to sustain in a controlled manner (hence the wishful hopes for finding "cold fusion"; see Chapter 10 for more on hydrogen). To put the power of the reaction into perspective, less than a gram (about a third of a packet of sugar) was all the matter that was converted when the Little Boy bomb was dropped on Hiroshima.

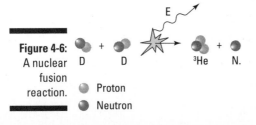

Figure 4-6:
A nuclear
fusion
reaction.

E

D   D ^{3}He   N.

○ Proton

● Neutron

Madame Curie

One of the most famous figures in modern science, Professor Marie Skłodowska-Curie is iconic for many reasons. She was born in Poland in 1867 and became very interested in Professor Henri Becquerel's radioactivity after he discovered it. This curiosity led her to isolate a new element, which she named Polonium after her native land. She later shared the 1903 Nobel Prize in Physics with Professor Henri Becquerel and her husband, Professor Pierre Curie. She was constantly researching nuclear physics with the hope of discovering new therapeutic uses for the materials. She also used the science during World War I to help diagnose suffering soldiers by using radium as an X-ray source for diagnostic imaging. Together with her daughter Irene, the two took X-rays of wounded soldiers to help find shrapnel to assist the surgeons. At night, the Curies would keep vials of glowing pitchblende in their room as a novelty (at the time the dangers of radiation were not known — little did they know their fancy nightlights were toxic). The Curies kept such close contact with radioactive elements that even today their lab notebooks give off a radioactive signature. She was not only a great pioneering scientist, but a pioneering woman in time that was ruled and governed by men. Fed up with the discrimination she felt, she eventually would hire only women or people who also experienced similar forms of discrimination. One such lady was Marguerite Catherine Perey, who started out in Curie's lab just washing test tubes. Perey later went on to discover the radioactive element Francium and was the first woman to be elected to the French Académie des Sciences.

Marie Curie went on to win a second Nobel Prize, this time in chemistry in 1911, for her discovery of radium and polonium. She was the first scientist noted for making the claim that radioactivity derives from within the atom and not by virtue of some unknown chemical reaction. This was the genius of Madame Curie, a woman immortalized through element 96: curium, with deep insight to the nature of the atom long before it's properties were revealed to the world.

Without fusion reactions, you would have a very boring periodic table of few elements (although it would make the test a little easier, it sure would be a boring world). The elements of Earth were originally synthesized in the stars like our own sun. Although our Sun only makes elements as high as iron, other more energetic stars can overcome the decreasing binding energy that occurs after iron and still make even heavier elements. These elements are then ejected into space in a supernova explosion and eventually end up at places like our little blue planet.

Chapter 5

The ABCs: Acid-Base Chemistry

⸱⸱⸱

In This Chapter

▶ Defining acids, bases, and the pH scale

▶ Reviewing how chemists have thought about acid-base chemistry

▶ Understanding relative characteristics of hard and soft acids and bases

▶ Moving beyond the pH scale with superacids

⸱⸱⸱

*A*cids and bases have been known for thousands of years due to their practical applications. To the ancient Greeks acids were referred to as *oxein,* meaning "sour tasting." In Latin they were known as *acere,* which came from the word *acetum,* meaning vinegar. Later, this was adapted to the English vocabulary as *acid.*

As these descriptive names suggest, acids were originally distinguished by their sharp, sour taste. Not only are they sour, but they were found to be corrosive to metals. Similar to rusting, acids oxidize metals. They were then found to react with lichen plants and caused litmus paper to change color.

Bases were often investigated and categorized according to how they neutralize, change, or counteract acid solutions. They were first recognized when making soaps using roasted ash. The word *alkaline* hails from the Arabic word "to roast." The soap-making process creates alkaline-water-based mixtures composed of both slaked lime and roasted ash.

Consider the expression "every reaction has an equal and opposite reaction." Consider acids and bases are each equal and opposite to each other; when you add together an acid and a base, they cancel each other out, and in that respect they are opposite to each other.

The terms *alkaline* and *base* are often used interchangeably, because they mean the same thing.

In this chapter, we explain how the scientific understanding of acids and bases has changed through time. We start by explaining how to measure their strengths using the pH scale. We also explain the multiple theories of acid-base chemistry, from Lavoisier's first attempt to classify these materials, to the modern understanding of Lewis acids and bases.

Starting with the Basics: Acids and Bases

In simplest terms, acids are ionic compounds that form hydrogen ions (e.g. HCl forms into H^+ and Cl^-) when dissolved in water; they have a characteristic sour taste and are corrosive. In contrast, non-acids are basic. They form negatively charged hydroxide ions (for example, NaOH forms into Na^+ and OH^-) when dissolved in water, and they are quite slippery to the feel. Bases are usually less sour tasting than acids (they tend to taste bitter), but they can be as corrosive as acids.

Although subjective observations can help you sort out some acids and bases, scientists have found they can quantify the strength of an acid or base by measuring the hydrogen ion concentration of a solution.

Developing the pH Scale

In 1889, Herman Walter Nernst measured the ion content in a solution as a function of the *electrode potential*. Shortly thereafter, a visual tool was developed by S. P. L. Sørensen who created a colorimetric assay. With this, he defined pH as the logarithmic concentration of the hydrogen ion. pH means the power of hydrogen and quantifies the power of hydrogen on a scale from 0 (very acidic) to 14 (very basic) based on the autodissociation of water. The pH of water is 7, lemon is 2.4, and bleach is 12.5, whereas beer is 4.5. Sørensen was working for a brewery at the time of his discovery, because brewing beer requires a careful balance of acid levels so that yeast and other materials can operate under ideal conditions. It was, therefore, necessary to develop a quantifiable measurement. The pH scale was largely ignored by the chemistry community, but was adopted by biological chemists because it was shown to have some practical importance for the life sciences and in other industries such as beer brewing. But its greatest impact was in the field of enzymology — the study of enzymes. Enzymes are large biological molecules that act as catalysts in the body (see Chapter 17), but they are very sensitive to changes in pH; therefore, by using the pH scale, a great deal of information could be gleaned for this avenue of science.

How pH became recognized

The measure of pH is important for many industries, and although the scientific community had not fully adopted it initially, it nevertheless gained ground for practical reasons. For example, the California Fruit Growers Association needed some way to measure acidity because it related to their citrus crops, so in 1934 they hired Arnold O. Beckman to invent an acidometer. Later, it became more and more evident that pH played a vital role in the manufacturing of oxide layers for the developing information technology industry.

Calculating pH

The pH value of an aqueous solution measures the H^+ concentration; this is measured in mol/L and is calculated using this equation:

$$pH = \log_{10} \frac{1}{[H^+]} = -\log_{10}[H^+]$$

But actually the H^+ ions are hydrated, so H^+ also means $H^+(H_2O)$, or H_3O^+ (this is called the hydronium ion). The autodissociation of water is described by this equation:

$$2\,H_2O \rightleftharpoons H_3O^+ + OH$$

Any time water is present, some of its ions are also present. At room temperature, the concentration of ions is known to be 0.00001%. This corresponds to a concentration of 1×10^{-7} M. Placing this value into the previous equation, shows that the pH of pure water is:

$$pH = \log_{10} \frac{1}{[1 \times 10^{-7}]} = -\log_{10} 10^7 = 7$$

Because the concentration of ions can be so cumbersome to write as percentage or as molarity (M) (see the glossary for a description of molarity), so the pH scale is used; it's a *log scale* that spans several orders of magnitude. Values less than 7 are acidic, and values higher than 7 are basic (1 M H^+ is pH 0, 10^{-14} M H^+ is pH 14).

Calculating acid dissociation

The *acid dissociation* can be calculated using the same method as the dissociation of water to determine the pH. The dissociation constant for an acid is the pK_a and is calculated this way:

$$pK_a = \log_{10} \frac{1}{K_a} = -\log_{10} K_a$$

When you see p in chemistry this means log_{10}. K_a is the equilibrium constant for the acid. Remember that the hydronium ion (H_3O^+) and H^+ proton can both form spontaneously in water; this is called autoionization. Therefore, there is some equilibrium concentration for the reaction where the hydronium ion is formed; this is defined using the symbol K_a because it's the equilibrium constant of the acid.

This is the measure of the strength of an acid — the stronger the acid, the more tendency to produce H^+ ions in solution. If it ionizes completely, then the K_a is large and the acid is strong. If it doesn't ionize very much, the pK_a is small.

Touring Key Theories: A Historical Perspective

Chemists have been studying and observing acid-base chemistry for hundreds of years. Over time, the nature of acids, bases, and how they react with one another has become more clear.

The early years

In the 1700s, Antoine Lavoisier postulated a formal and theoretical classification system for acids and bases. He made two assumptions — that all acids contained an essence (which he suggested was oxygen), and that all acids were similar with no unique differences among them. He was wrong on both counts, but his assertions engaged the scientific community nonetheless.

In 1884 Arrhenius proposed his theory of electrolytic dissociation based on the self-ionization of water:

$$H_2O \rightleftharpoons H^+ + OH$$

When H^+ was created, the substance was called an acid, and when OH^- was created the substance was called a base. Where a typical neutralization reaction is:

$$HCl\ (aq) + NaOH\ (aq) \rightarrow NaCl\ (s) + H_2O$$
$$(acid)\ +\ (base)\ =\ (salt)\ +\ (water)$$

(acid) + (base) = (salt) + (water). Humphrey Davy challenged the prevailing notion of oxygen as the essence of an acid and instead suggested that hydrogen was the key. He developed his ideas while working with hydrochloric acid, HCl, that he found to contain no oxygen.

Following Davy's work, two chemists, Justus Von Liebig and Svante August Arrhenius, continued to explore acid-base chemistry. They set forth three assertions:

- An acid delivers hydrogen; in other words, it donates H atoms into a water solution.

- Bases deliver hydroxyl groups, OH, by donating them to the water solution.

- Acid-base reactions neutralize one another, and the product of a balanced reaction will always render a salt product and water.

Brønsted-Lowry theory

In 1923, two independent scientists, Johannes Brønsted and Thomas Lowry, defined acids to be proton donors and bases to be proton acceptors. Continuing to work with the concept of hydrogen in acids and bases, Brønsted and Lowry each independently defined a new set of conditions that has come to be known as the *Brønsted-Lowry model*.

They continued with the notion of acids being proton donors, but they expanded on the concept of the base to include any substance that is capable of binding protons. In other words, any compound that complexes with a Brønsted acid is considered a Brønsted base.

In aqueous solutions this is not a major distinction, but it was known that other acid base reactions can occur in solvents that are not water. The Arrhenius theory applies to water. But as with the water example, where the self-ionization of water occurs, hydronium ion concentration increases above a value of 10^{-7} M for acids, and those that decrease the value are bases.

The Brønsted-Lowry theory expands on this notion beyond solvents of just water, to cover others such as liquid ammonia, glacial acetic acid, anhydrous sulfuric acid, and all solvents containing hydrogen. Bases can accept protons but don't necessarily contain OH^-.

For example, in liquid ammonia, the NH_4^+ donates a proton, whereas the NH_2^- accepts a proton.

$$NH_4Cl + NaNH_2 = Na + Cl - + 2NH_3$$
$$(acid) + (base) = (salt) + (solvent)$$

When the chemical species differ in composition by a proton, they are called a *conjugate pair*. For every acid there is a conjugate base, and vice versa.

$$A\,(acid)\quad B\quad (conjugate\ base) + H+$$
$$B\,(base) + H+\quad A + (conjugate\ acid)$$

In water this looks like:

Acid — Conjugate base

$$HCl + H_2O \rightleftharpoons H_3O^+ + Cl^-$$

Base — Conjugate acid

In this reaction, the HCl is an acid because it donates protons, which forms free Cl^- as a base.

In liquid ammonia, this looks like:

Acid — Conjugate base

$$NH_4^+ + S^{2-} \rightleftharpoons HS^- + NH_3$$

Base — Conjugate acid

In this reaction, the ammonium salt acts as the acid, because it donates the protons, whereas the sulfide ion is a base because it accepts the protons. This reaction is reversible and proceeds to produce the species that is weakest, such as HS^- and NH_3.

REMEMBER

This explanation of acids and bases depends on the choice of solvent. If a solvent is chosen that is a strong donator of protons, then it is considered the acid in the reaction, even if it's not technically called an acid. Or, if two acids are mixed together, the weaker donor of protons is instead classified as the base. Take, for example, the case of hydrogen fluoride mixed in perchloric acid:

$$HClO_4 + HF\quad H_2F + + ClO_4 -$$

Both of these materials can be individually considered acids, but when they are mixed together the hydrogen fluoride gains a proton and the perchloric acid gave one up. In this case, the perchloric acid was a stronger acid because it was the acid that donated the H^+, and the HF was a weaker acid, so it acted like a base and accepted the H^+.

Accepting or donating: Lewis's theory

Up to this time, the ideas about acids and bases focused on the use of water as a solvent. Water is a protic solvent because it has the capacity to donate protons. Water is polar, which means it has a slight electric charge (see Chapter 10 for more details). Gilbert Lewis considered the case of a nonpolar, or nonprotic, solvent and how dissolution might occur in non-aqueous solvents where polar species cannot be dissolved.

Lewis basically threw out the two previous ideas and instead looked at the reactions based on the transfer of electron pairs. He postulated that acids accept electron pairs, and bases donate electron pairs.

In the Lewis model, an acid accepts electron pairs, whereas a base donates electron pairs. Any proton is a Lewis acid, whereas ammonia, for example, is a Lewis base, because the lone pairs of the nitrogen are donated to a proton as shown in the following equation. (In this example the backward pointing arrow is used to signify that the electrons are being donated.)

$$H^+ + :NH_3 \rightarrow [H \leftarrow :NH_3]^+$$

Organic chemists commonly refer to Lewis acids as *electrophiles* and Lewis bases as *nucleophiles*. A classic example of a Lewis acid-base reaction involves BF_3 and NH_3. The ammonia has a lone pair that can donate to the empty p orbital of the BF_3. In this reaction, the BF_3 is the Lewis acid and NH_3 is the Lewis base. This results in the formation of a covalent bond between them.

The Lewis theory is most useful when looking at the reactions of the main group elements and transition metal ions.

The importance of the Lewis concept of acids and bases is that it can help explain reactions between molecules that are not strictly acids or bases, so it broadens the applicability to include nearly every reaction you can think of.

The concept that donation is not restricted to acids gives chemists the possibility to apply these same simple rules to more and more complicated systems, but still make sense of them because they follow these simple trends. One molecule gives something up, and the other molecule takes it on.

Comparing Lewis and Brønsted theories

Because you can look at acids as being either proton donors or as electron acceptors (and bases vice versa), you commonly see a complex labeled as either a Lewis acid/base, or Brønsted acid/base. For this reason it's valuable to consider the similarities and differences between these definitions.

Here are the key points to remember:

- Brønsted dealt only with acid-base chemistry in water.

- In the Brønsted model remember that an acid is a proton donor, and a base is a proton acceptor.

- Brønsted: Acid/base reaction B: + H-A $\rightarrow$ B$^+$-H + A$^-$. In this case, the base accepts the proton from the acid.

- pH = $-\log_{10}$[H$^+$] comes from the Brønsted definition of acid in water.

- Brønsted base and Lewis base are related by the fact that both have the ability to complex a proton.

- Brønsted acid is a H$^+$ donor, while at the same time H$^+$ is itself a Lewis acid. So, it's also true that any compound that can complex a proton is both a Brønsted base and a Lewis acid.

- H$^+$ is a unique Lewis acid; it readily takes an electron to balance the positive charge. When the proton switches back and forth, it's considered a Brønsted-type reaction.

- In the Lewis model an Acid is an electron acceptor, and the Base is an electron donor.

- Lewis: Acid/base reaction B: + A $\rightarrow$ B$^+$ − A$^-$. The electron is donated to the acid from the base.

- Lewis acid interacts by the lowest unoccupied molecular orbital (LUMO), Lewis base interacts via the highest occupied molecular orbital (HOMO). (See Chapter 6 for more details about HOMO/LUMO.)

- Lewis base is anything that can complex a Lewis acid. This is a broad classification that could consist of nucleophiles, ligands, spectator ions, or e-rich π-systems.

- To analyze Lewis acid/bases-type reactions, watch where the electrons go in the reaction.

Pearson's Hard and Soft Acids and Bases (HSAB)

Using the ideas set out by Lewis that acids and bases are either electron acceptors or electron donors, a professor of inorganic chemistry, Prof. Ralph G. Pearson (currently at UC Santa Barbara) noted similarities in the ways that acids and bases react with one another, particularly in the case where two competing acids might try react with a base. He thought about why one acid was more likely to react with a base over some other kind of acid. He wondered what property made it more likely to react and what trends or general rules could help predict this process. Based on the way that a series of acids were grouped together, he noted that "hard" acids tend to prefer "hard" bases, while "soft" acids tend prefer "soft" bases.

Take for example the reaction of two acids, hard and soft respectively, A_1 and A_2, reacting with two bases, B_1 and B_2, that are soft and hard respectively:

$$A1B1 + A2B2 = A1B2 + A2B1$$

The reaction proceeds in the direction where a hard acid bonds to a hard base, and a soft acid bonds with a soft base.

The idea of hard and soft isn't based on a quantifiable value; it's a comparative test between various acids and various bases. For one reaction an acid is considered hard, but in the presence of a more hard acid it's the softer of the two and is considered the soft acid. It's like how people judge the weather; for example a cold day in Texas may be considered a warm summery day in Michigan.

However, if the bonding between an acid and base is more covalent in nature (see Chapter 6), then they are considered soft. Atoms with high charge densities, low oxidation states, low electronegativities, and a high capacity for the electron orbitals to become polarized (see Chapter 8) lead to covalent bonding.

The elements themselves can also be considered hard or soft, depending on the oxidation state. This is true for example with nickel. Ni(0) is a soft acid and is stabilized by soft bases such as CO and $Ni(CO)_4$. But for example, Ni(V) is a hard acid and is instead stabilized by hard bases like oxide ions, such as NiO_4^{3-}. When it's in a high oxidation state, it's considered a hard acid, so it prefers to complex with a hard base.

One drawback of Pearson's hard and soft acids theory is that there is no scale for relative hardness or softness. With two hard acids, one may be harder than the other, but there is (so far) no way to quantify or measure that difference.

Pearson's theory is very useful for organic chemistry because it can be used to explain electrophilic and nucleophilic substitution reactions that are very common organic chemistry reactions.

Characterization of the hard bodies

To determine whether an acid or base is hard or soft, you need to look at what kind of bonding will occur between them. If the acid-base complex forms according to an ionic bonding structure (see Chapter 8), they are hard. Properties that lead to ionic bonding include a high-charge density and a tendency to undergo electrostatic interactions.

Charge density gives a measure of how much and how tightly packed the charge is around an atom. According to Fajan's rule, the charge density of a cation affects the extent of covalent bonding. See Chapter 11 for more details on Fajan's rule.

High charge density is a good property to judge a material as a hard acid. For example, H^+, B^{3+}, and C^{4+} act like hard acids. Electrostatic interactions between acids and bases are usually caused by the high charge density of the atoms. They tend to form ionic type bonds between them because of the high electrostatic nature of the atoms. High charge density materials include the majority of metal ions in the periodic table H^+. As well as cations with high charge densities and very low electronegativity, such as metals Mg^{2+}, and nonmetal fluorides BF_3.

With respect to bases, the hardness increases with the group number of the donor atoms, for example $NH_3 < H_2O < F^-$. While at the same time hardness decreases going down the table: $F^- > Cl^- > Br^- > I^-$. This includes fluorine and oxygen-bonded species such as oxides, hydroxides, carbonates, nitrates, phosphates, perchlorate, and sulfates. Monoatomic ions have high charge densities and are also considered hard bases.

Who you callin' soft?

Soft acids and bases typically have low charge densities, coupled with high electronegativity, resulting in readily polarized orbitals. Because the orbitals can be polarized, their shape can morph a little bit so it can bond with other soft species. This happens because of the orbital interactions between each atom and how each has the capacity for polarization of the orbital — and remember that when this happens, the bonds tend to form as a covalent type bond.

Soft acids are electron pair acceptors (also called Lewis acids). The accepting atom has a zero or low positive charge, and are relatively large in size. This leads to low electronegativity and valence electrons with high polarizability, which makes them easy to oxidize.

Soft acids include cations of the late transition metals and post-transition metals such as Cu^+, Pd^{2+}, and Hg^{2+}. Gold(I) is the softest. Soft bases are electron pair donors (Lewis bases). The donor atom has high polarizability and can exhibit pi-acceptor behavior, such as CO. Typically soft bases are nonmetallic atoms with low electronegativity. Examples include carbon, sulfur, iodine, and phosphorus.

Strapping on a Cape: Superacids

In 1927 James Bryant Conant created an acid that was 1 million times stronger than any acid previously made, and for this reason he coined the term "super-acids." He achieved this by the addition of sulfuric acid and fluorosulfuric acid. Then in 1960 George Olah created a "magic acid" that was strong enough to dissolve candle wax, a first of its kind — this wasn't possible to do before, and was achieved by mixing antimony pentafluoride (SbF_6) with fluorosulfuric acid (HSO_3F).

Superacids are extremely strong acids that are usually found in solutions other than water. In water, the strongest acid that can exist is H_3O^+; trying to make a stronger acid in water just protonates H_2O to make it H_3O^+. Superacids, however, are 10^6 (1 Million) to 10^{10} (ten Giga) times more acidic than high concentrations of strong acids.

Measuring super acids requires something beyond the normal pH scale from 0 – 14. The superacid scale, or *Hammett acidity function*, is based on experimental evidence. For example, in this equation

$$H_0 = pK_{BH^+} - \log_{10} \frac{[BH^+]}{[B]}$$

H_0 is the Hammett acidity function, B is an indicator base, BH^+ is in the protonated form, and pK_{BH+} is the $-\log K$ for the dissociation of BH^+. The ratio $[BH^+]/[B]$ is measured using spectrophotometry. Spectrophotometry is a technique that uses light to measure the properties of matter.

Using bases with very low basicities (negative pK value), the H_0 scale is extended to very negative values. The H_0 scale can be used in the same way that the pH scale can be used. In essence, it can be thought of as the pH scale for values less than pH=0.

Part II
Rules of Attraction: Chemical Bonding

The 5th Wave   By Rich Tennant

At The Local Chemists' Watering Hole

Whoa! Look at the pocket protectors on this one!

In this part . . .

Bonding between atoms is an important concept for all scientists to understand because it affects the properties and applications of materials in profound ways. There are hundreds of different atoms from the periodic table to bond with, but there are only three types of bonding known — covalent bonding, ionic bonding, and metallic bonding. In this part, you learn how bonding is carried out and how to distinguish the types of bonding that are used. You also learn how to describe the shapes and structures that these molecules can make.

Chapter 6

No Mr. Bond, I Expect You to π: Covalent Bonding

....

....

*W*hen covalently bound atoms run home to show their mom their report card, they can always brag about their excellent marks in "Plays well with others." Covalent bonds (perhaps better written as *co-valent* bonds) are formed when two or more atoms share their valence electrons.

This chapter walks you through the evolution of covalent bond theory, starting with the Lewis dot structures you likely covered in General Chemistry and then advancing to valence bond and molecular orbital theories that stem from quantum mechanics calculations. As with most chemistry, this chapter is purely about following the bouncing balls (electrons).

Connecting the Dots: Lewis Structures

Remember the head football coach with a whistle 'round his neck who always had his whiteboard in hand on game day? He's going to be our model example of how to draw pictures that represent atoms known as Lewis structures (and a model example of what not to wear on a Friday night).

The first thing that coach would draw would be a little symbol for each of the players in the game. Similar to our coach's Xs and Os, most electrons in Lewis structures are drawn with an *x* or a dot (·). In this book, I use dots.

Going with new numbers

There is a possibility that the periodic table you are looking at has different group numbers going across the top of it than the one shown in Figure 1-1. If, instead of numbers 1 through 18, you see Roman numerals with As and Bs, you have an old periodic table. The International and American standards varied slightly in the past, so the new 1–18 standard was designed to get rid of the confusion. Simply cross through the I over hydrogen and write 1, and continue across the periodic table until you place an 18 over helium.

Now, if you were a benchwarmer like me, there was never a symbol on the board for you; only the players who actually play in the game earn one of coach's Xs. The same is true when drawing electrons on a Lewis structures: Only the electrons that can participate in chemical bonding, the *valence* electrons, earn a spot on the Lewis structure drawings. The electrons that are in the inner shells of an atom don't play a role in bonding, so they are ignored.

For an easy way to calculate the number of valence electrons an atom has, simply look to the element's Group Number on the periodic table:

- For Groups 1 through 12 (the s- and d-blocks), the Group Number equals the number of valence electrons.

- For Groups 13 through 18 (the p-block), simply subtract 10 from the Group Number to determine the valence electrons.

Figure 6-1 illustrates some sample Lewis structures.

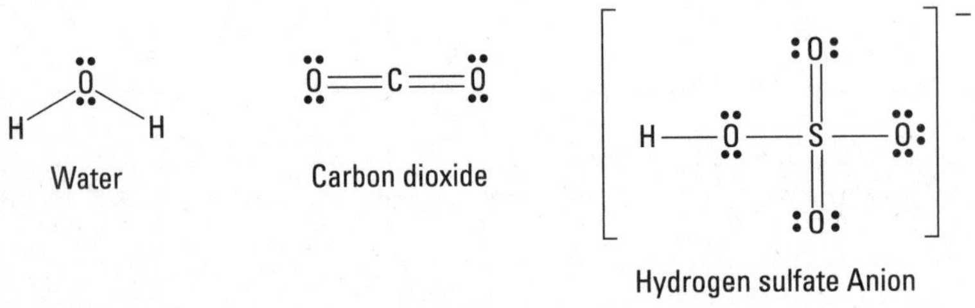

Figure 6-1: Examples of Lewis structures.

Water

Carbon dioxide

Hydrogen sulfate Anion

Counting electrons

When drawing Lewis structures, your goal is to fill the outer shell of electrons and make a *noble gas configuration*. For our coach, this is an easy task: He needs to show 11 players on the field. Any more than 11 is a foul; any less, and his team has a disadvantage. For atoms, the number of valence electrons to achieve noble gas configuration changes depending on where you are on the periodic table:

✔ **Duet rule:** With only a $1s$ orbital capable of holding two electrons ($1s^2$), helium is simply too small to hold eight electrons like the rest of the noble gases. Likewise, hydrogen ($1s^1$) wants just one electron to achieve $1s^2$, just as lithium ($1s^2\,2s^1$) wants to rid itself of the $2s^1$ electron to become $1s^2$.

Note: These atoms are still following noble gas configurations; this rule simply accounts for the noble gas (He) of the first row holding only two electrons.

✔ **Octet rule:** For the remainder of the main-group elements, the octet rule dictates the number of valence electrons an atom wants. The eight electrons are directly related to the s^2p^6 noble gas electronic configuration. With two electrons in the s orbital and six electrons in the p orbitals, the noble gas is full with eight electrons (hence the term *octet* rule).

✔ **18-electron rule:** This rule applies for the transition metals, specifically the Cr, Mn, Fe, and Co triads. Adding the five d-orbitals (each with two electrons) to the s^2p^6 of the noble gases, our rule goes up from 8 to 18. Atoms that follow the 18-electron rule are less likely to exchange coordinating ligands (ions or molecules that bind to it) and be less reactive, as I discuss in Chapter 15.

While this 18-electron rule holds up for many of the transition metals, a common exception is seen with atoms that have a d^8 electron configuration. These atoms commonly prefer a 16-electron configuration and assume square-planar molecular geometry, as illustrated in Figure 6-2.

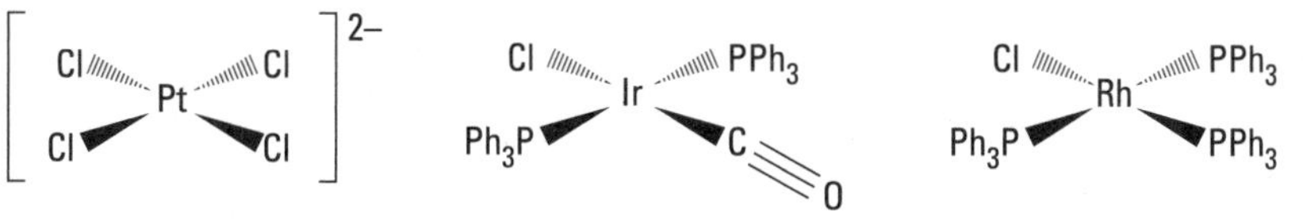

Figure 6-2:
Some 16-electron exceptions: $[PtCl_4]^{2-}$, Vaska's Complex, and Wilkinson's catalyst.

Electron counting works for most atoms, yet special cases exist where atoms have less or more than the noble gas configuration you're aiming for. Atoms with less electrons, commonly seen with boron and aluminum, are known as *electron-deficient*. Atoms with more electrons, such as certain sulphur, phosphorus, silicon, iodine, and xenon molecules, are known as *hypervalent* or *hypercoordinated.* Another common exception is seen with free radicals, which have a lone electron that requires the total electron count to be odd (so it clearly cannot follow the duet, octet, or 18 electron rule).

Placing electrons

Now that you've counted how many electrons you have in the game, it's time to make your play. In the following steps, you make a Lewis structure for xenon oxytetrafluoride ($XeOF_4$).

1. **Determine the total number of valence electrons.** Xe is in Group 18 (eight valence electrons), O is in Group 16 (six valence electrons), and F is in Group 17 (seven valence electrons each). This makes the total number of valence electrons 42. Nonvalence electrons are not used in Lewis structures.

2. **Place your atoms.** Usually, the least electronegative atom goes in the center of your Lewis structure (with the exception of hydrogen, which is never the central atom); then add the surrounding atoms. Place one pair of electrons between each pair of atoms. (Remember: a bond is two electrons.) Keep in mind that xenon, being a main group element beyond the second row of the periodic table, can be *hypervalent* (or have more than eight valence electrons).

3. **Place the remaining valence electrons (in pairs).** The best way to start is by placing the electrons around the most electronegative atom until you satisfy its noble gas configuration. Having used ten electrons in Step 2 for bonds, you have only 32 remaining valence electrons to place. Fluorine is the most electronegative. With two electrons currently around it (from the bonds), each F needs three more electron pairs to have noble gas configuration.

   With those 24 electrons placed, only 8 remain. Six go to oxygen to satisfy the octet rule, and the last two go to Xe.

4. **Place your bonds.** Fluorine wants one bond with three lone pairs, whereas oxygen wants two bonds with two lone pairs. This step is explained in more detail in the upcoming section "Price tags in black ties? Formal charges."

Here are some good general rules to follow when double-checking your final Lewis structures:

- ✔ Hydrogen and halogens want one bond.
- ✔ Oxygen wants one or two bonds.
- ✔ Nitrogen wants three bonds.
- ✔ Carbon wants four bonds.

Atoms[3]

Before scientists had a good grasp on the quantum mechanics that defined atomic orbitals, there was some confusion as to how to explain some of the periodicity of the elements on the periodic table. The definitions of valence electrons have changed over the years, but at one point the term resembled our current view of an oxidation state (you could have negative and positive valences). German chemist Richard Abegg noted that for any element, the most absolute values of the most negative valence and most positive valence almost always summed up to 8. Knowing what we know now, this is obvious. The valence shell can hold eight electrons, so if you take away all the valence electrons or fill the valence electrons, you always get a sum of the filled valence shell.

In 1902, Gilbert N. Lewis (yes, the same Lewis of our "Lewis dot structures") suggested a teaching tool using cubes to explain this strange trend. Imagine an eight-cornered cube on which you add one electron to each corner as you go across the periodic table from Li to Ne. Therefore, carbon would be a cube with four corners with dark spheres representing the four valence electrons. A single bond would occur when two atoms shared one edge (or two electrons), and a double bond would occur when two atoms shared a face of the cube (or four electrons). As you can see, this isn't too far from the truth. Unfortunately for Lewis, the theory got a little hazy when trying to include triple bonds, yet it undoubtedly led to his development of the octet rule we still use today.

Price tags in black ties? Formal charges

Well, which resonance structure should you pick? To help answer this question, scientists came up with a snazzy little equation that tells you the charge on an atom if all the electrons in all the bonds of the molecule are shared equally.

Bonds aren't shared equally if they have *electronegativity* differences, or differences in how much an atom tends to attract the electrons around it. But, for now, let's play along with the scientists' equation.

A formal charge is determined by the following equation:

$$FC = VE - LPE - \tfrac{1}{2}BE$$

FC is the formal charge, VE is the number of valence electrons, LPE is the number of lone pair electrons, and BE are the number of bonding electrons. You take half of the bonding electrons because you are assuming that every bond is shared 100 percent equally, or each atom gets one electron of each bond.

If (like me) you are allergic to equations, you will really like the circle method. On your Lewis structures, simply draw a circle around the atom you want to know the formal charge of. Make sure that circle includes the atom's lone pairs, and split all the bonds on that atom in half. Now the easy part: Add 'em up. Simply count the number of electrons in the circle and treat each split bond as one electron. These are the total electrons around the atom for the purposes of a formal charge. The last step is to simply subtract the number you got from the atom's formal charge (remember, just look to the group number). Voila! A formal charge. Figure 6-3 shows an example of both the equation and the circle method for the nitrate anion (NO_3^-).

Oxygen ($2s^2\ 2p^4$) has 6 valence e⁻s
Nitrogen ($2s^2\ 2p^3$) has 5 valence e⁻s

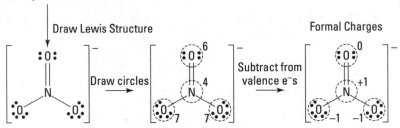

Figure 6-3: Determining the formal charge on a nitrate anion (NO_3^-).

Upper oxygen has 2 lone pairs and 2 half bonds (6 e⁻s)
The nitrogen has 4 half bonds (4 e⁻s)
Lower oxygens have 3 lone pairs and 1 half bond (7 e⁻s)

The formal charge represents how much the atom changes: the atom starts with a certain number of electrons (the valence electrons) and finishes with a certain number of electrons (the number of lone pairs and bonds). Nature, being lazy as she is, prefers to spend the least amount of energy — she wants to avoid change. Knowing this, the structures with formal charges closest to zero, or the lowest change from start to finish, are the most stable. Minimizing your formal charges helps you choose the dominate resonance structures.

Sometimes you get a molecule with resonance structures that both have low formal charges yet have the negative formal charge on different atoms like the cyanate anion ([NCO]⁻) in Figure 6-4. Two of the three resonance structures have formal charges of 0s with one (–1) on either the nitrogen or oxygen atom. Simply reducing the formal charges doesn't help you pick which one is correct. A positive (+) formal charge means that the atom lost an electron by forming the molecule, whereas a negative (–) formal charge means that the atom gained an electron. Knowing this, you want to assign the negative formal charge to the most electronegative atom.

Figure 6-4:
Placing the
negative for-
mal charge
on the most
electronega-
tive atom (O).

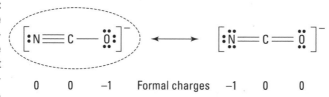

0   0   −1   Formal charges   −1   0   0

Returning to the drawing board: Resonance structures

You may notice that some Lewis structures with double bonds can be drawn multiple ways. For example, the molecule of ozone (O_3) shown in Figure 6-5 can be drawn with the double bond on the left oxygen atom or the right oxygen atom.

Figure 6-5:
Drawing
an ozone
molecule.

When this situation occurs, these two structures are said to be *resonance structures* of one another. When resonance occurs, this is a sign that the electrons of the molecule are *delocalized* or not associated with one single bond in the molecule.

When electrons become delocalized, it's important to remember that all the resonant Lewis structures are just partial pictures of how the molecule may appear. Just as the ozone molecule can be drawn with the double bond left or right, it can more correctly be drawn with a partial double bond to both oxygen atoms as shown on the right.

Some texts may refer to resonance structures as *canonical forms* or *contributing forms*. Regardless of what you call the resonant Lewis structures, remember that the structures are *not* isomers. With isomers, you actually change the structure of the molecule. For a resonance structure, you effectively pin the atoms to the paper and only move electrons around throughout the molecule.

Keeping Your Distance: VSEPR

Have you ever watched two siblings on a road trip? No matter how close they may be, by the end of that trip, they are physically as far apart from one another in the car as possible.

Pairs of electrons, childish as they are, behave in a similar manner — those negative charges want to be as far apart as possible. The *valence shell electron pair repulsion* (VSEPR) theory is used to predict the shapes of molecules based on the idea that similarly charged electron pairs want to be as far apart from one another as possible. Though it looks like V-S-E-P-R, the theory name is most commonly pronounced "vesper."

Figure 6-6 shows how the pairs of electrons arrange around a central atom (and stay on their side of the car). Note that a *steric number* is the number of atoms and lone pairs bonded to the central atom of a molecule.

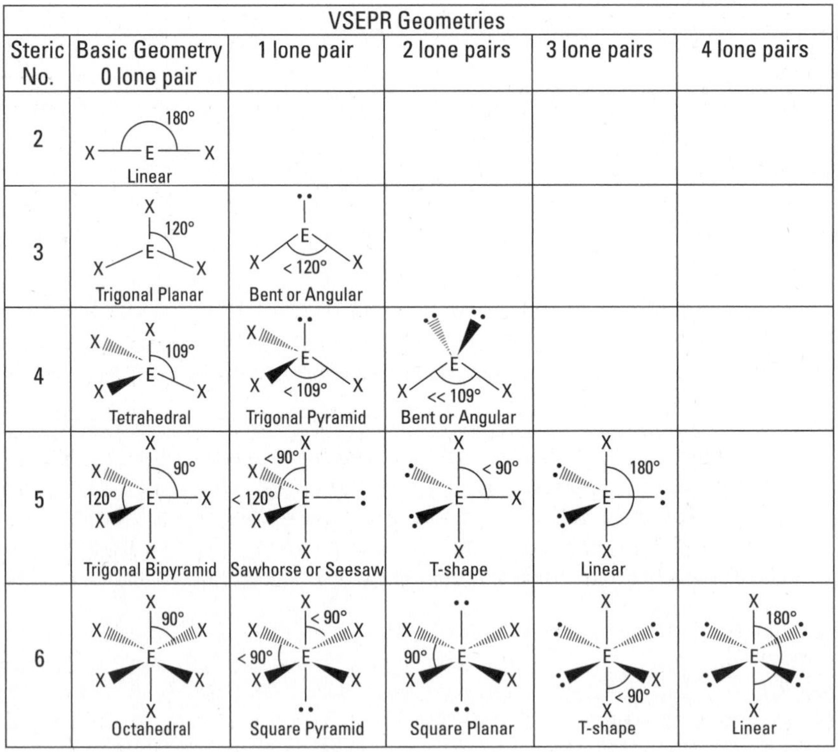

	VSEPR Geometries				
Steric No.	Basic Geometry 0 lone pair	1 lone pair	2 lone pairs	3 lone pairs	4 lone pairs
2	180° Linear				
3	120° Trigonal Planar	< 120° Bent or Angular			
4	109° Tetrahedral	< 109° Trigonal Pyramid	<< 109° Bent or Angular		
5	120° 90° Trigonal Bipyramid	< 120° < 90° Sawhorse or Seesaw	< 90° T-shape	180° Linear	
6	90° Octahedral	< 90° < 90° Square Pyramid	90° Square Planar	< 90° T-shape	180° Linear

Figure 6-6: VSEPR geometries.

With two or three pairs of electrons, the resulting arrangement of electrons is flat. However, when four or more pairs of electrons are around the central atom, the electrons arrange themselves in a 3D manner:

- **Four electron pairs:** The electrons form the peaks of a triangular pyramid called a *tetrahedron*. They're similar to the pyramids in Egypt, but with a triangular base. (For you Dungeons and Dragons fanatics, the shape is like a four-sided die.)

- **Five electron pairs:** Start with the trigonal planar shape that exists with three electron pairs, and then envision sticking a skewer straight through the sheet of paper.

- **Six electron pairs:** You get an *octahedron* (take two four-sided pyramids and glue their square bases together) where all the electron pairs are 90 degrees apart.

When you replace electron pairs with bound atoms, you actually make molecules, and VSEPR can help you figure out what the molecules look like. For example, when you draw the Lewis structure for water, you get two lone pairs of electrons and two bound atoms attached to O (for a total of four "things," or a steric number of four). Figure 6-6 shows that an atom with four things bound to it creates a tetrahedron, but when two of those "things" are invisible, you're left with a bent molecule (or a molecule that looks like a boomerang).

Figure 6-6 covers the vast majority of molecular shapes you will encounter. Molecules with higher steric numbers, such as the uranium (V) fluoride ion (UF_7^{2-}), are incredibly rare.

With just electrons, the *bond angles* (the angles between two pairs of electrons) are idealized. If you substitute all the electrons with identical atoms, the same effect will be seen. For example, if C is the central atom and you add four H atoms (I just got you to make methane and you didn't even need to eat beans), each H atom will be 109.5° apart, as expected (see Figure 6-7). However, if you pick N as your central atom with three H atoms around it (ammonia, NH_3), the bond angle decreases to 107.1°. Moreover, when you get to H_2O with two electron pairs, the bond angle between the H atoms decreases to 104.4°. As a rule, the amount of repulsion will be:

2 electron pairs > an electron pair to a bound atom > 2 bound atoms

Figure 6-7:
Bond angles for methane, ammonia, and water.

When comparing different bound atoms, look to electronegativity for help as to what will repel more. The higher the electronegativity, the more negative charge near the bound atom, and thus more repulsion.

Ante Up One Electron: Valence-Bond Theory

Lewis drawings give you pictures of electron pairings between atoms; valence-bond (VB) theory goes beyond the pictures and adds the quantum mechanics we all love so much (huzzah for page-long equations!). Essentially, this theory hinges on the idea that atomic orbitals overlap, as shown in Figure 6-8.

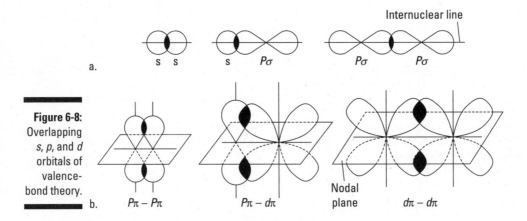

Figure 6-8:
Overlapping *s*, *p*, and *d* orbitals of valence-bond theory.

Notice that when the p-orbitals come together, you can have either one lobe from each p-orbital involved in head-on overlapping (a sigma [σ] bond) or you can have both lobes overlapping (a pi [π] bond). In molecules, single bonds have one sigma bond, double bonds have one sigma bond and one pi bond, and triple bonds have one sigma bond and two pi bonds.

Sigma bonds allow for rotation, whereas the pi bonds do not. For the sigma bond, you could twist the p-orbitals in opposite directions and never have to break the bond. However, if you try to twist the p-orbitals in opposite directions in a pi bond, you would have to break the bonds.

Linus Pauling was the mastermind of valence-bond theory. To explain how overlapping orbitals form a bond, he came up with a set of rules to describe what happens when two atoms share a pair of electrons to form a bond:

> ✔ Each bond forms from an unpaired electron of each atom.
>
> ✔ The spins of the electrons have to be opposite (one ↑ and one ↓).
>
> ✔ When paired (↑↓), the two electrons cannot take part in other bonds.
>
> ✔ The orbitals with the most overlap form the strongest bonds.

These rules work wonderfully for overlapping *s* orbitals, like in molecular hydrogen (H_2). However, if you go back to methane, things act strangely. Carbon, the central atom, has an electron configuration of $1s^2 2s^2 2p^2$. Remember that the *p* orbitals are the three bow-tie-looking orbitals that go along the *x*, *y*, and *z* axes. The 1s orbitals of the H atoms want to form the strongest bonds possible, yet overlapping the three *p* axes along the *x*, *y*, and *z* axes places the H 90° apart. However, VSEPR suggests that methane wants to be a tetrahedron with the H 109.5° apart.

To account for this problem, Pauling proposed the idea that atomic orbitals can come together and form *hybridized* orbitals, or mixed orbitals, that will conform to the shapes you expect. For example, in Figure 6-9 you can see how one s-orbital (ball shaped) and three p-orbitals (bow-tie shaped) can come together to form four funny-shaped bow-ties that are 109.5° apart as you'd expect for anything with a steric number of 4.

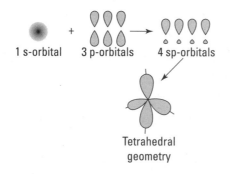

1 s-orbital   3 p-orbitals   4 sp-orbitals

Tetrahedral
geometry

s-orbital   p-orbital   Hybridize

Two *sp*   *sp* hybrid orbitals
hybrid orbitals   shown together
(large lobes only)

Figure 6-9:
Hybridizing
orbitals.

2s $2p_x + 2p_y$ 3 sp² orbitals

With these new, shiny hybridized orbitals, atomic orbitals allow for bonding as expected (and follow the expected VSEPR models of Figure 6-6). Each hybridized orbital simply combines the orbitals its name implies. So an sp orbital is a combination of an s orbital and a p orbital. The number of orbitals hybridized always equal the number of hybrid orbitals created. For example (assuming no lone pairs):

- ✔ An sp orbital has two hybrid orbitals resulting in a linear molecule.

- ✔ An sp^2 orbital has three hybrid orbitals resulting in a trigonal planar molecule.

- ✔ An sp^3 orbital has four hybrid orbitals resulting in a tetrahedral molecule.

- ✔ An sp^3d orbital has five hybrid orbitals resulting in a trigonal bipyramidal molecule.

- ✔ An sp^3d^2 orbital has six hybrid orbitals resulting in an octohedral molecule.

Summing It All Up: Molecular Orbital Theory

Valence-bond theory does a good job of describing bonding where the electrons are localized into atomic orbitals. A counter theory involves the elimination of atomic orbitals for the creation of molecular orbitals — hence the name, molecular orbital (MO) theory.

An atomic orbital is just a region where you can expect to find an electron that stems from an equation known as the *Schrödinger equation.* If you add two atomic orbitals together, you get a region of overlap where the electrons will likely be found between the atoms. Similarly, if you subtract them, you will likely find the electrons in regions away from the area between the atoms as shown in Figure 6-10. These sums and differences are often referred to as the *linear combination of atomic orbitals* (or LCAO).

Types of MOs

Given the choice, a negatively charged electron would much rather hang out between two positively charged nuclei (opposites attract, you know). Because of this fact, the constructive (or in-phase) overlap in Figure 6-10 is lower in energy (electrons are lazy and want to be in the lowest energy state possible).

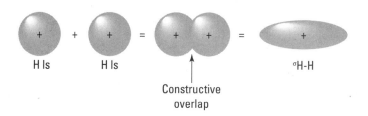

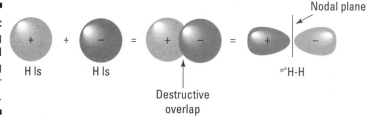

Figure 6-10: Forming bonding and antibonding molecular orbitals.

There are three different types of MOs:

- **Bonding MOs:** These MOs are constructive interactions that are lower in energy than the atomic orbitals they started in before bonding. These are denoted as the type of bond (σ, π, and so on).

- **Nonbonding MOs:** These MOs occur when the atomic orbitals can't really come together to bond, so the electrons stay in the same energy before and after bonding. They do not aid or hamper bonding and are confined to a single atom (think of lone pairs). These are denoted as the type of bond with a NB superscript (σ^{NB}, π^{NB}, and so on).

- **Antibonding MOs:** These MOs are the deconstructive interactions that are higher in energy than the atomic orbitals they started in before bonding. These are denoted as the type of bond with an asterisk (σ^*, π^*, and so on).

Evens and odds: Gerade and ungerade symmetry

When orbitals overlap in σ and π bonds, they can do so in a symmetric or asymmetric manner. If you invert everything through a point directly between the atoms, if the resulting orbitals look exactly like what you started with, you have an even (or *gerade*) bond. If they look opposite, you have an uneven (or *ungerade*) bond. You designate these MOs with a subscript of g or u, respectively. For example, an even π bond would be π_g. Figure 6-11 shows examples of gerade and ungerade π and σ bonds.

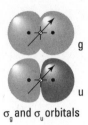

σ_g and σ_u orbitals

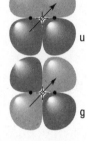

Figure 6-11:
Gerade and ungerade σ and π bonds.

π_g and π_u orbitals

Identical twins: Homonuclear diatomic molecules

To begin examining this theory, first look at a simple *homonuclear diatomic* molecule: molecular hydrogen (H_2). Each hydrogen has a single 1s orbital with one electron dancing around in there. When you bring two hydrogens together (as shown in Figure 6-10), they overlap where the electrons are shared between the two nuclei (a σ bond) or where the electrons are not between the two nuclei (a σ^* [pronounced *sigma-star*] bond). The star designates that it is an antibonding orbital. The negatively charged electrons would want to be closer to the positively charged nuclei if given the chance, so the σ^* orbital isn't the most welcoming to electrons; given the chance, they'd rather stay as atoms than bond in that molecular orbital — hence the term *anti*bonding.

To form these molecular orbitals, the atomic orbitals that are mixing must be close in energy. For each two atomic orbitals that mix, a bonding and an antibonding orbital will be formed. The antibonding is always higher in energy than the bonding orbital. These newly formed molecular orbitals follow similar rules to atomic orbitals:

✔ Only two electrons are allowed per orbital, and they must have opposite spins (one ↑ and one ↓).

✔ If you have two molecular orbitals equal in energy, you must fill them all with one electron before you start pairing them (↑ ↑ not ↑↓ ___).

✔ The spins will want to be parallel if in different orbitals (↑ ↓ not ↑ ↑).

Figure 6-12 shows molecular orbital energy diagrams (the higher up on the diagram, the higher the energy of the electron) for diatomic molecules.

Note: He_2^+ looks like the right scheme, but is missing one electron in sigma*.

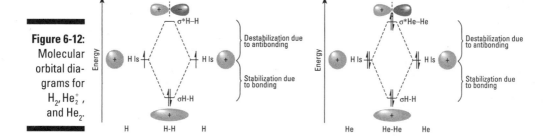

Figure 6-12: Molecular orbital diagrams for H_2, He_2^+, and He_2.

Bond order

In reality, diatomic He doesn't exist. *Wait a second, I just saw the molecular orbital diagram for it in Figure 6-12!* Well, there's a term known as *bond order* that tells you how strong a bond should be. Simply subtract the number of antibonding electrons from the bonding electrons, divide by 2, and, voila!, you have the bond order. Moreover, the number corresponds to the number of chemical bonds between atoms (a single bond is 1, a double bond is 2, and so on). So, from Figure 6-12, H_2 has a bond order of 1, He_2^+ has a bond order of 0.5, and He_2 has a bond order of 0. Therefore, there is no strength in the bond.

Here's another way of thinking about this: The bonding orbitals are lower in energy than the atomic orbitals, which is appealing to those lazy electrons. The antibonding orbitals, however, are more like a day in the gym. Well, when days on the couch eating ice cream equal the days at the gym, there's no net benefit. Same here. The electrons that dropped in energy are equal to the electrons that increased, so why bother bonding?

HOMO/LUMO

You may hear the outermost electron-occupied orbitals of molecules referred to as *frontier orbitals*. Because the outermost electron-occupied orbital is also the highest in energy, you often hear it referred to as the *highest occupied molecular orbital* (HOMO). And in case you're wondering (and I'm sure you are), the *lowest unoccupied molecular orbital* is called the LUMO.

Figure 6-13 is the energy diagram that comes after bringing the p orbitals of oxygen together to form one sigma and two pi bonds (and their corresponding antibonding orbitals) and the s orbitals together to form one sigma bonding and antibonding orbital. It's interesting to note that molecular oxygen is *paramagnetic:* It has two unpaired electrons. You can see this experimentally by pouring liquid oxygen over a magnet and watching it stick to the magnet.

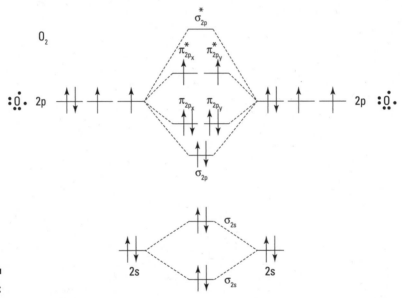

Figure 6-13:
Molecular
orbital dia-
gram for O_2.

Electron Configuration: $(\sigma_{2s})^2 (\sigma_{2s}*)^2 (\sigma_{2p})^2 (\pi_{2p})^4 (\pi_{2p}*)^2$

Bond Order $= \frac{1}{2}(2 - 2 + 2 + 4 - 2) = 2$ Double Bond

Fraternal twins: Heteronuclear diatomic molecules

Next, you get a little more intricate in the details because the atomic orbitals in a *heteronuclear diatomic* (two atoms of different nuclei) molecule are starting in different energies. Unlike the homonuclear diatomic molecules, these molecules are *polar* (the electrons will favor one atom to the other). The electrons are going to naturally want to hang out with the more electronegative atom. Because of this, when you are building MOs with two different atoms, the more electronegative atom will contribute more to bonding orbitals, while the less electronegative atom will contribute more to the antibonding orbitals. Figure 6-14 shows a molecular orbital diagram for a molecule of CO.

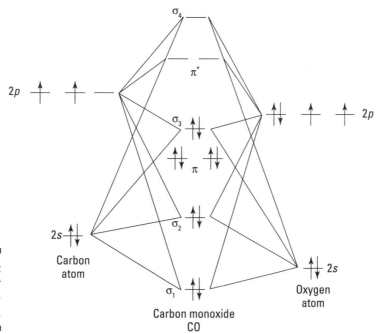

Figure 6-14:
Molecular orbital diagram for CO.

Chapter 7

Molecular Symmetry and Group Theory

. . . .

In This Chapter

▶ Spying on symmetry elements and operations

▶ Mastering molecular point groups

▶ Figuring out a character table

. . . .

Symmetry, or the property of having balanced proportions about a plane or axis, is all around us. The examples are common — a monarch butterfly; the Taj Mahal; snowflakes. Molecules also have symmetry and molecules with similar *molecular symmetry* are often grouped together as they often exhibit similar properties. From *spectroscopy* (hitting your chemicals with energy and waiting to see how they respond) to *crystallography* (looking at how atoms are arranged in a solid), molecular symmetry is incredibly useful for determining properties of molecules, such as polarity, infrared activity, Raman activity, chirality, and more. This chapter explains how molecular symmetry can be used to assign a molecule to a three dimensional group, known as a *point group*, that other molecules with the same symmetry elements can be assigned; moreover, each molecular point group has a corresponding table, known as a *character table*, that's incredibly useful in predicting and determining properties of the molecule using spectroscopy methods, such as infrared and Raman spectroscopy.

Identifying Molecules: Symmetry Elements and Operations

The three-dimensional symmetry of a molecule can be described by elements and operations:

✔ A *symmetry element* is the point of reference (a point in space, a plane, or an axis, for example) around which the symmetry occurs.

✔ A *symmetry operation* is the rearrangement of atoms around a symmetry element.

Take Alice from *Alice in Wonderland* as an example. Imagine Alice holding her looking glass (mirror). Everything on Alice's side of the mirror looks like everything inside the mirror (Alice looks exactly like Alice's reflection), so the plane of the mirror is acting as a symmetry element. However, when Alice magically steps through the mirror, Alice becomes Alice's reflection (and Alice's reflection becomes Alice). The swapping of Alice and her reflection about the plane of the mirror (a symmetry element) is a symmetry operation.

A molecule can possess five symmetry elements, about which symmetry operations can be performed:

✔ Identity (E)

✔ *n*-fold rotational axis (C_n)

✔ Inversion center (*i*)

✔ Mirror plane (σ [yet another sigma . . .*sigh*])

✔ Improper rotation axis (S_n)

In this section, I walk you through each of them.

Identity

The first symmetry element is simply an *identity* denoted as E. (The E stems from the German word for unity.) As the main title of this section suggests, every element has one E that simply says the molecule exists, so even the most lopsided, ugliest molecule you can imagine has at least one E. Because "you are what you are," the operation about the identity is to do absolutely nothing — you cannot change the molecule's identity (and there's no identity theft in the molecular world).

Establishing a term to describe doing nothing may seem silly at first, but E is useful for expressing that a molecule is back to where it started when performing multiple symmetry elements. Imagine performing the Alice-through-the-looking-glass switcheroo twice. The first time, Alice becomes Alice's reflection (and vice versa). The second time, Alice returns to being Alice (and her reflection to being the reflection); we can say Alice has returned to E.

n-fold rotational axis

Determining an *n-fold rotational axis* (C_n) is like sticking a skewer through an object and twisting the skewer. Imagine you are at a party and the host brings out a plate of hors d'oeuvres (greatest point in the party, right?). On the plate, you instantly notice the scrumptious cocktail wieners begging to be eaten. As you carefully select your wiener of choice, you stab the wiener straight through the middle with a toothpick. You then roll the toothpick between your fingers. After you rotate the toothpick 180°, the left half of the wiener becomes the right half (and vice versa). Back on your hunt-and-gather mission, you notice a mini-sandwich cut into a triangle. Again, after stabbing your grub through the middle, you notice you have to rotate the toothpick only 120° for the sandwich to look the same as how it started. Finally, you decide to stab a meatball. This time, even after the tiniest twitch, the meatball still looks like how it started.

At this point, you, the professional food-twirler that you have become, have performed a C_2, C_3, and $C\infty$ rotation, respectively. To determine the n for C_n, you simply take the number of degrees in a circle (360°) divided by the number of degrees you twisted the molecule around the rotation axis to make the molecule look like it started. For example, for the cocktail wiener, 360° /180° = 2, whereas for the sandwich, 360° /120° = 3. Because the meatball is a sphere, any rotation, no matter how large or small, looks like the start, so the value of n is infinity.

Many molecules have more than one rotational axis. For example, back at the cocktail party, instead of stabbing the sandwich through the center from atop, stab it through one of the triangular tips from the side. Again, you have skewered the sandwich, but now when you twist the toothpick, the sandwich twirls in a different direction. This time, you have to twist the 180° to have it resemble the starting point (C_2). In fact, you could do this through any of the three triangular tips. At this point, your sandwich has one C_3 and three C_2s. The *n-fold rotational axis* with the highest value of n is defined as the *principal axis*. When drawing molecules, you always want to place the principal axis up-and-down (or along the z-axis).

One final point is a note on keeping track of your rotations. If you rotate a molecule along a C_n for n times, you return to the molecule's identity (E). To keep track of how many times you have rotated the molecule, you add a superscript to denote how many times you have performed the C_n. Figure 7-1 demonstrates this process on BF_3, the molecular version of our cocktail party sandwich. Notice how after three C_3s, you're back to the original molecule.

In 3D drawings of molecules:

- ✔ Solid lines means the bond is in the plane of the paper.

- ✔ Solid triangles means the bond is coming towards you (out of the paper).

- ✔ Broken triangles means the bond is going away from you (into the paper).

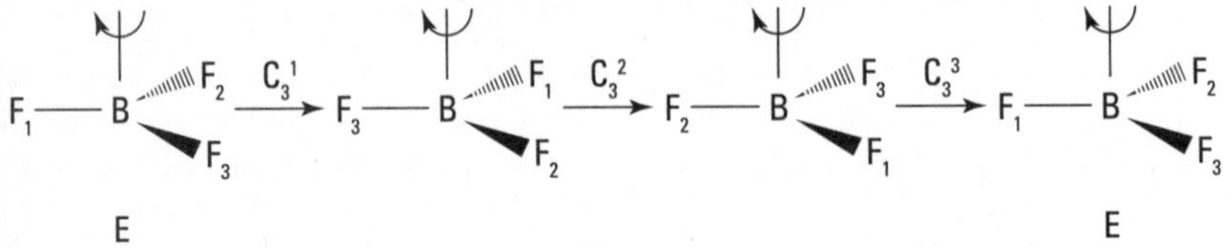

Figure 7-1:
A three-fold rotational axis on boron trifluoride (BF_3).

Inversion center

The next symmetry element is called an *inversion center* (*i*). Simply put, place a point (*i*) at the center of your molecule and pull everything straight through it to the other side. If it appears the same after the inversion, it's an inversion center. You may also hear this point referred to as a *center of inversion* or a *center of symmetry*. Figure 7-2 shows how BF_3 does not have an inversion center, although XeF_4 does.

Think of this process as turning an object inside out. Consider a reversible sock with an interior exactly like the exterior. Turning it inside out, performing an inversion, reverses the sock. Now consider a reversible leg warmer. An inversion produces the original (E).

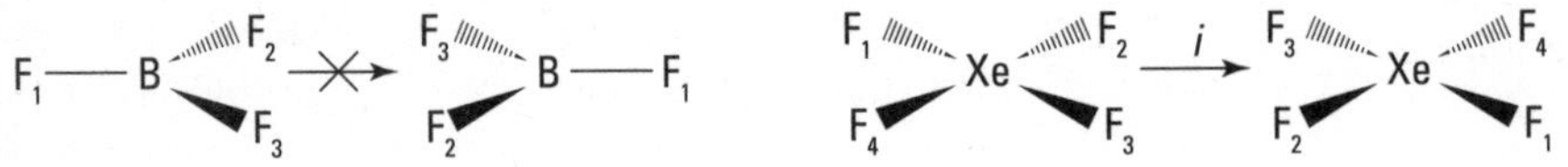

Figure 7-2:
Inversion centers (*i*).

Mirror planes

So now return to Alice's looking glass. Quite literally, she was holding a *mirror plane* (σ) — a plane in space that sliced her worlds (the real world and the reflection world) into equal halves. If you can slice a molecule at any point and make everything on one side look like the other side, you have a mirror plane. There are three types of mirror planes:

✔ **Vertical mirror planes (σ_v):** These are mirror planes that are, as you might guess, vertical. The key here is making sure you have the principal rotation axis along the z-axis; otherwise, it won't be vertical.

✔ **Dihedral mirror planes (σ_d):** These mirror planes exist if you can create a mirror plane between two C_2 rotation axes. XeF_4 in Figure 7-3 has a σ_d between the two C_2 axes (one C_2 going vertically through F_1-F_3 and the other going through F_2-F_4).

✔ **Horizontal mirror planes (σ_h):** These mirror planes are perpendicular to the principal rotation axis. Figure 7-3 shows how BF_3 has a σ_v whereas $XeOF_4$ does not.

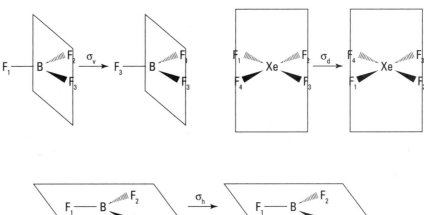

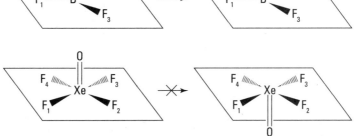

Figure 7-3:
Vertical,
dihedral,
and horizon-
tal mirror
planes for
BF_3, XeF_4,
and $XeOF_4$.

Improper rotation axis

The final symmetry operation, an improper rotation (S_n, the S coming from the German word *spiegal* for mirror) is the most complex as it is a combination of two previously defined symmetry operations: a reflection through a mirror plane, and an *n*-fold rotation. This is one of those cases where a picture is worth a thousand words. Figure 7-4 shows an improper rotation on a molecule of methane (CH_4). ***Note:*** I rotated the molecule away from the principal axis going along the z-axis to make it easier to see.

First, you perform a C_4 (90°) rotation on the methane molecule. Notice this rotation is not symmetric (the molecule looks different following the C_4 rotation from where it started); however, after we reflect the image through a mirror plane, we are back to a molecule similar to where we started. Huzzah!

Chirality

A molecule is *chiral* if it cannot be superimposed on its mirror image. Imagine that you are holding an invisible magic ball in your hands with your palms facing up (or, if you have an invisible magic ball, go get it real quick). Look at your hands. Your left hand is the mirror image of your right hand. However, if you try to overlap your hands so they superimpose, you see they are clearly different. When your palms are up, your left thumb points left and your right thumb points right. To be superimposable, they would need to point the same direction.

The definition for a chiral molecule is that it simply cannot possess an S_n operation. (But remember: $S_1 = \sigma$ and $S_2 = i$, so you often see it written that a chiral molecule can't possess a σ, i, or S_n.)

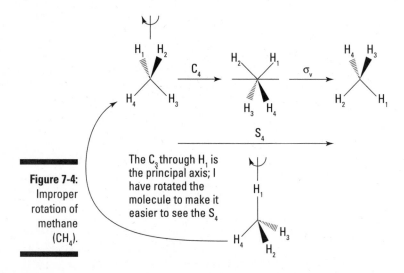

Figure 7-4:
Improper rotation of methane (CH_4).

The C_3 through H_1 is the principal axis; I have rotated the molecule to make it easier to see the S_4

An S_1 would mean that you rotate a molecule 360° (back to where it started) and then reflect, so $S_1 = \sigma$. An S_2 would mean that you rotate a molecule 180° and then reflect, so you get everything opposite from the center from where it started (which is identical to an i, so $S_2 = i$). Consider a baseball with red stitching: a 90-degree rotation followed by a reflection through the vertical mirror plane produces the original ball.

It's Not Polite to Point! Molecular Point Groups

Just because elements have the same molecular structure doesn't mean they have the same symmetry. For example, if you take the methane molecule from Figure 7-4 and sum up all the symmetry elements, you find four C_3 and three C_2 axes, as well as six σ_v planes. If you swap all the hydrogen atoms with chlorine atoms (CCl_4 or carbon tetrachloride), your new molecule has the exact same symmetry. However, if you swap only three of the hydrogen atoms with chlorine atoms ($CHCl_3$ or chloroform), your symmetry decreases because the number of symmetry elements decrease in number ($CHCl_3$ has only one C_3 axis and only three σ_v planes). So even though CH_4, CCl_4, and $CHCl_3$ are all tetrahedral in shape, only CH_4 and CCl_4 have the same symmetry.

Because both CH_4 and CCl_4 have the exact same symmetry elements about a single point (the middle of the C atom), we classify them as being in the same *point group*. Some molecules have incredibly low symmetry, such as bromochlorofluoromethane (CHBrClF has only E), methanol (CH_3OH has only E and σ_h), and 1,2-dichloro-1,2-difluoroethane ($C_2H_2Cl_2F_2$ has only E and i), as shown in Figure 7-5. These molecules belong to the low symmetry point groups of C_1, C_s, and C_i, respectively.

Figure 7-5:
Molecules
of low
symmetry.

Bromochlorofluoromethane
C_1

Methanol
C_s

1, 2-dichloro-1, 2-difluoroethane
C_i

Similarly, some molecules belong to point groups of high symmetry. Examples include methane (CH_4, tetrahedral [T_d] point group), molybdenum hexacarbonyl ($Mo(CO)_6$, octahedral [O_h] point group), and buckminsterfullerene (C_{60}, icosahedral [I_h] point group) as shown in Figure 7-6.

Linear molecules also have high symmetry because they have a C_∞ rotation axis going through the molecule, as shown for carbon monoxide and nitrogen in Figure 7-7. Note that carbon monoxide's point group is a $C\infty_v$ whereas nitrogen's point group is $C\infty_h$. The difference from C (for cyclic) and D (for dihedral) stems from the fact that nitrogen has a C_2 axis perpendicular to the primary rotation axis (in the case of nitrogen, the $\check{C}\infty$).

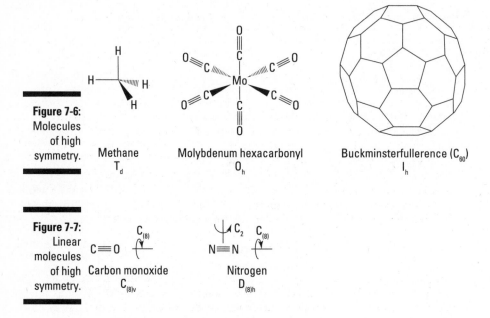

Figure 7-6:
Molecules
of high
symmetry.

Methane
T_d

Molybdenum hexacarbonyl
O_h

Buckminsterfullerence (C_{60})
I_h

Figure 7-7:
Linear
molecules
of high
symmetry.

Carbon monoxide
$C_{(8)v}$

Nitrogen
$D_{(8)h}$

To assign a molecule to a specific point group outside of these high and low symmetry point groups, you need to either follow this list of steps or go through the point group flow chart of Figure 7-8. Both methods determine the same result.

For point group determination:

1. **Determine if your molecule has low symmetry (C_1, C_s, C_i), high symmetry (T_d, O_h, I_h), or is linear ($C\infty_v$ or $D\infty_v$).**

2. **If not one of these groups, determine the principal rotation axis (the C_n with the highest order or largest n).** The n determined for the principal rotation axis will be the n in all the following subscripts.

3. **Determine if there is a C_2 rotation axis perpendicular ($\perp$) to the principal rotation axis.** If so, you have a D group — go to Step 4. If not, you have either a C group or an S group — go to Step 6.

D point groups

4. **Determine if your molecule in the D point group has a σ_h.** If so, you have a D_{nh}. If not, go to Step 5.

5. **Determine if your molecule has a σ_d.** If so, you have a D_{nd}. If not, you have a D_n.

C or S point groups

6. **Determine if your molecule in the C or S point group has a σ_h.** If so, you have a C_{nh}. If not, go to Step 7.

7. **Determine if your molecule has a σ_v.** If so, you have a C_{nv}. If not, go to Step 8.

8. **Determine if your molecule has an S_{2n}.** If so, you have an S_{2n}. If not, you have a C_n.

There is no such thing as a D_{nv}. If you know your molecule is D ($C_2 \perp$ to the principal rotation axis), and you find a vertical mirror plane, look closer at your plane — it's truly dihedral (bisecting two C_2).

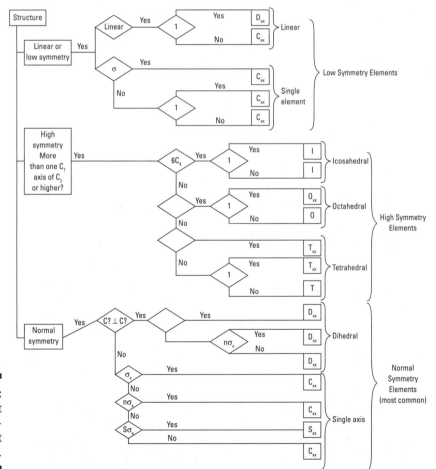

Figure 7-8:
Flow chart to determine point groups.

Being Such a Character Table

If you flip to the back of most inorganic textbooks, one of the appendices will have tons of tables in it. Unlike my appendix, which doctors removed a few years back, the character table appendix is quite critical. These *character tables* display (in what looks to be cryptic hieroglyphics at first glance) fundamental information on how symmetry operations affect things of interest (orbitals, bonds, atoms, and so on) within a specific point group. Figure 7-9 displays a character table for the C_{2v} point group (the same point group as a molecule of water).

Table 7-9:
Character
table for C_{2v}
point group.

C_{2v}	E	C_2	$\sigma v(xz)$	$\sigma v^2(yz)$		
A_1	1	1	1	1	z	x^2, y^2, z^2
A_2	1	1	-1	-1	R_z	xy
B_1	1	-1	1	-1	x, R_y	xz
B_2	1	-1	-1	1	y, R_x	yz

In this section, I explain how to navigate through a character table in order to glean the information you need.

Dissecting a character table

The character table in Figure 7-9 has six distinct sections:

- ✔ The **upper-left corner** has the desired symmetry point group (C_{2v}).

- ✔ The **top row** lists all the symmetry elements in the point group divided into *classes*. The number of classes is simply the number of columns in that segment of the character table. For C_{2v}, there are four classes: E, C_2, σ_v (xz), and σ_v^2(yz).

The *order* (h) of a character table is the sum of all the symmetry elements. For C_{2v}, again the order is 4; however, the number of classes and the order of the matrix are not always the same. Sometimes, the symmetry elements have a *coefficient* in front. The coefficient simply means there are multiple equivalents of that symmetry element included in that class. For example, there are six symmetry elements for the point group C_{3v}: E, C_3, C_3^2, σ_v, σ_v', and σ_v''. Because C_3 and C_3^2 have identical columns in the character table, however, you find them grouped together into one class on your character table with a 2 in front of C_3 (as in: $2C_3$). You find the three mirror planes grouped into one class as well, so you are left with three classes on the C_{3v} character table (E, $2C_3$, and $3\sigma_v$) but an order of 6.

If a C_2 is listed as C_2', the prime is telling you that the C_2 is perpendicular ($\perp$) to the principal axis and the C_2 is going *through* atoms. Likewise, if the C_2 is listed as C_2'', it is a C_2 perpendicular ($\perp$) to the principal axis going *between* atoms. If you look back to the XeF_4 molecule in Figure 7-2 before the inversion, C_2' would be the C_2 axis going through F_1, Xe, and F_3, while C_2'' would be the C_2 axis going between F_1 and F_2, through Xe, and between F_3 and F_4.

✔ The **far-left column** with the letters and subscripts presents the *Mulliken symbols,* which simply identify the *irreducible representations* (all the numbers in the row, which I describe in further detail in the next bullet). For C_{2v}, you find A_1, A_2, B_1, and B_2. Here's how to interpret what you're seeing:

- *Letters:* These tell you about the number of *dimensions* (also called the *degeneracy*) of the representation (row of numbers). The dimension of a representation is simply the value under *E* on the character table. Because all molecules have an identity operation (*E*), *E* is always listed as the first class of symmetry elements (the first column of numbers). For C_{2v}, the dimensions of all four representations are 1 (or, every row has a value of 1 under *E*), so you would expect the letter of the Mulliken symbol for each representation to be either an A or a B:

    A: One dimensional (*E* = 1) and symmetric when rotating around the principal axis.

    B: One dimensional (*E* = 1) and asymmetric when rotating around the principal axis.

    E: Two dimensional (*E* = 2).

    T: Three dimensional (*E* = 3). **Note:** In older character tables, you may find that an F is used instead of a T. (In fact, a G equals quadruple degeneracy, and an H equals quintuple degeneracy!)

    The sum of the squares of all the dimensions always equals the character table's order. For C_{2v}, adding up the square of all the numbers under the E class renders 4: $(1)^2 + (1)^2 + (1)^2 + (1)^2 = 4$.

- *Subscripts and superscripts:* These tell you more about the symmetry of the representations with regards to a symmetry element (similar to A and B with regards to the principal axis):

    Subscript *g* or *u* (for *gerade* and *ungerade*): Symmetric or asymmetric following inversion (*i*), respectively.

    Subscript 1 or 2: If there is a C_2 perpendicular ($\perp$) to the principal axis, these subscripts represent symmetry and asymmetry to that C_2, respectively. If there is no perpendicular C_2, these subscripts represent symmetry and asymmetry about a vertical plane (σ_v), respectively. **Note:** Higher numbers as subscripts mean there are additional representations with asymmetry.

Prime (') or double prime ("): Symmetry or asymmetry about a horizontal plane (σ_h), respectively. You find these on molecules with both a horizontal plane and an odd n for the principal axis (C_s, C_{3h}, D_{5h}, and so on).

✔ The **rows of numbers in the middle** of the character table are the *irreducible representations,* which I discuss in greater detail later in this chapter. At this point, know three things:

- Every character table has a completely symmetrical representation (every value is 1); C_{2v} has A_1.

- Every representation on the character table is completely *orthogonal* (the dot product is zero). The best way to check for orthogonality is to take two representations, multiply each class together, and sum up the products. For example, A_2 is orthogonal to B_1:

$$E \qquad C_2 \qquad \sigma_v(xz) \quad \sigma_v'(yz)$$
$$(1)(1) + (1)(-1) + (-1)(1) + (-1)(-1) = 0$$

You get zero no matter what two irreducible representations you choose.

- The sum of the squares of any irreducible representation always equal the character table's order (h). For example, C_{2v} has an order of 4 and the sum of the squares of B_2 is:

$$E \quad C_2 \quad \sigma_v(xz) \quad \sigma_v'(yz)$$
$$(1)^2 + (-1)^2 + (-1)^2 + (1)^2 = 4$$

✔ The **second-to-last column** of the character table lists the linear functions and rotations associated with each irreducible representation. These functions describe the symmetry transformations of the three Cartesian coordinates (x, y, and z) and rotations about those three coordinates (R_x, R_y, and R_z).

✔ The **far-right column** of the character table lists the quadratic functions (or functions involving the quadratic terms of x^2, y^2, z^2, xy, xz, or yz) associated with each irreducible representation that are helpful for both IR and Raman spectroscopy, as discussed later in this chapter.

If you see two functions from lower right columns inside parenthesis separated by a comma, such as (R_x, R_y), then the functions are doubly degenerate (they appear in an irreducible representation that begins with the letter E and will have a value of 2 in the E class). Conversely, if the two functions are separated by a comma without parenthesis, like z, R_z, then both functions contribute to that irreducible representation independent of one another.

Degrees of freedom

One way of describing three-dimensional space is to describe it in terms of x, y, and z (known as *Cartesian coordinates*). When you use these coordinates to describe a molecule, you can then refer to the three types of motion that a molecule can have: translational modes (moving back and forth), rotational modes (think about spinning the molecule), and vibrational modes (this is everything else). Because these motions are ways that a molecule is free to move, they are referred to as *degrees of freedom*. For any molecule, there are $3N$ total degrees of freedom if the molecule has N atoms.

Always align the principal axis (C_n with the largest n) along the *z-axis*.

✔ Translational modes are movements that change the position of the molecule (think of grabbing a molecule and sliding it along a line). All molecules have three translational modes of freedom: along the x-, y-, or *z-axis*.

✔ Rotational modes are movements that change the position of the molecule (think of grabbing a molecule and spinning it about an axis). Most molecules have three translational modes of freedom: along the x-, y-, or *z-axis*. There are two major exceptions to this rule: monoatomic species (think noble gases) and linear molecules. For the lone atom, rotation along the x-, y-, or *z-axis* doesn't change anything, so there are no rotational modes of freedom. (Also, there are only three degrees of freedom when N = 1, and the monoatomic species already has three translational modes.) For the linear molecule, there is no change when you spin the molecule along the *z-axis*.

✔ Vibrational modes are all the other internal movements a molecule possesses. For the translational and rotational modes, all the atoms were moving together in sync; however, for vibrational modes, the atoms are moving relative to one another. For example, water has three atoms. Because it is not linear or monoatomic, water has three translational modes and three rotational modes. Because water has nine degrees of freedom (3N for N=3), there are three remaining vibrational modes.

Because this discussion is all about movement, think of a gym in the late 1980s. Undoubtedly, someone was sweating out to a workout on Suzanne Sommers' ThighMaster. One of water's vibrational modes bends just like the ThighMaster. (Never seen one before? Google it!) Imagine the joint being the oxygen atom and the tips of the two metal loops being your hydrogen atoms. Water, just like the ThighMaster, bends back and forth.

Similarly, imagine someone working out with elastic resistance bands (those strong elastic bands with barbell grips on the ends — in the 1980s, they were likely neon pink and yellow). The athlete steps on the middle of the band, grabs the two handles, and begins to lift with all his might to stretch both bands up toward the ceiling. If you imagine the athlete's foot (scratch that, don't think about athlete's foot . . . let's say athlete's shoe) as being the oxygen atom, the two handles would become hydrogen atoms. Because both hydrogen atoms are stretching away from the oxygen symmetrically, we call this type of vibration a *symmetrical stretch*.

However, imagine that the athlete decides to quit lifting the handles to the ceiling and instead begins doing biceps curls one at a time. The left arm curls up and lengthens the band while the right arm curls down and shortens the band. This form of stretching would be *asymmetric* and also constitutes the third and final vibrational mode you would find in water.

To quickly determine the number of vibrational modes, keep this hint in mind:

- Monoatomic: 0
- Linear molecules: $3N - 5$
- Nonlinear molecules: $3N - 6$

A glitch in the matrix: Matrix math

Despite your hopeful wishing, the *character* of a character table does not involve Mickey Mouse. Instead, this term is a way of describing a square matrix (and no, not a simulated reality in the future where Keanu Reeves has been replaced with Ben Stein). A *matrix* is simply a rectangular arrangement of numbers or symbols. That rectangular matrix is defined by having m rows and n columns, making it an mxn (pronounced: m-by-n) matrix. Therefore, [1] is a 1x1 matrix as it has 1 row and 1 column, while [1 2 3] is a 1x3 matrix. A square matrix simply has m equal to n.

When multiplying matrices together, the order of the matrices being multiplied matters: The first matrix must have the same number of columns (m) as the second matrix has rows (n). This means that you can multiply a 1x2 matrix followed by a 3x1 matrix in that order, but not a 3x1 matrix followed by a 1x2 matrix.

To multiply two matrices, multiply the numbers of the same row (n) of the first matrix by the numbers of the same column (m) from the second matrix and add them all up. For all cases, make sure you are multiplying in the row n of the first matrix equal to the column m of the second matrix. Here are the steps:

1. **Multiply the first number of a row (n) from the first matrix with the first number of the same column (m) from the second matrix.**

2. **Add your result to the second number of the row (*n*) from the first matrix times the second number of that column (*m*) from the second matrix.**

3. **Continue the same process until you multiply the last number of the row (*n*) from the first matrix by the last number of that column (*m*) from the second matrix and add it to the sum.**

Examples of matrix multiplication are shown in Figure 7-10.

Figure 7-10:
Examples of
matrix multi-
plication.

$$\begin{bmatrix} a & b \\ c & d \end{bmatrix} \begin{bmatrix} e & f \\ g & h \end{bmatrix} = \begin{bmatrix} ae + bg & af + bh \\ ce + dg & cf + dh \end{bmatrix}$$

$$\begin{bmatrix} 1 & 2 \\ 3 & 4 \end{bmatrix} \times \begin{bmatrix} 5 & 6 \\ 7 & 8 \end{bmatrix} = \begin{bmatrix} (1)(5)+(2)(7) & (1)(6)+(2)(8) \\ (3)(5)+(4)(7) & (3)(6)+(4)(8) \end{bmatrix} = \begin{bmatrix} 19 & 22 \\ 43 & 50 \end{bmatrix}$$

Conveniently, three Cartesian coordinates (*x*, *y*, and *z*) describe where all the atoms of the molecule can be found. Mathematically, you can manipulate these coordinates to move the atoms. For example, think of a one-dimensional line (just an *x*-axis). Place a big circle on the line at any positive number (call it *x*). Now perform an identity (*E*) operation on the point: Nothing changes. Your point is still at *x*. Now perform an inversion through the center (0) of your number line: Your circle moves from *x* to –*x*. So for this point on your line, the *E* operation was the same as multiplying by 1 (nothing changed) and the *i* operation was the same as multiplying by –1 (it moved to the other side of zero on the line).

When you switch from 1D to 3D, you use matrix multiplication in the same manner. Multiplying your circle's position by 1 or –1 transforms its position, and multiplying your atom's position in 3D coordinates by a *transformation matrix* does the same. A transformation matrix can be used to multiply by a matrix of 3D coordinates (*x*, *y*, and *z*) to determine their new location relative to the starting position. Every symmetry operation can be described as a transformation matrix.

Figure 7-11 shows four molecules of water with arrows on each atom in the *x*, *y*, and *z* directions (these conveniently would overlap with the lobes of p_x, p_y, or p_z orbitals) following the four symmetry operations associated with a molecule in the C_{2v} point group. Remember, the identity operation (*E*) doesn't change anything, so you can use *E* as a starting point for the other operations. Notice that when you perform a C_2, the H_1 and H_2 switch locations and their *x* and *y* arrows change directions. Under each symmetry operation, you find a 3x3 transformation matrix and the results of transforming the original coordinates (*x*, *y*, and *z*).

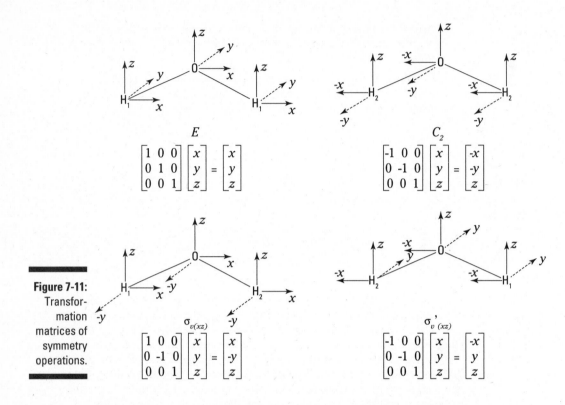

Figure 7-11:
Transfor-
mation
matrices of
symmetry
operations.

Now, back to what started this whole section: the character! The character is the sum of the values in a square matrix if you go diagonally from the upper-left to the lower-right. For the four transformation matrices of water:

E	C_2	$\sigma_v(xz)$	$\sigma_v'(yz)$
3	-1	1	1

The set of characters is known as a *reducible representation* (designated Γ) because it can be composed of the irreducible representations that make up a character table. In this case, the reducible representation is the sum of A_1, B_1, and B_2 from the character table, and notice in Figure 7-12 how the linear functions of x, y, and z correspond to three irreducible representations found in the C_{2v} character table.

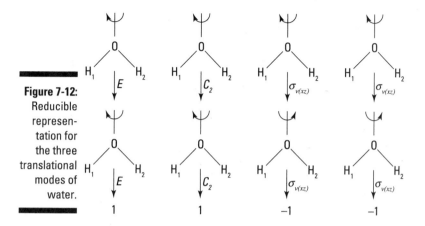

Figure 7-12: Reducible representation for the three translational modes of water.

Reducible reps

When looking for a reducible representation, simply think of a molecule with three vectors (x, y, and z) on every atom, like the water molecule in Figure 7-11. Now, as before when developing your transformation matrices, you need to see what happens to the vectors when doing each symmetry operation. However, there's one key difference: You care only about atoms that do not move locations. If an atom appears somewhere different after performing the operation, do not count any of the three vectors.

For water, the E operation keeps all nine vectors in the same direction because nothing changes. Performing a C_2 causes the two hydrogen atoms to switch locations, so you ignore those six vectors. Oxygen's z-vector stays the same, so you assign it a value of 1; oxygen's x- and y-vectors switch directions so you assign them a value of –1. With one 1 and two –1s, the total for the operation becomes –1. For the σ_v (xz) mirror plane, you see that all three atoms stay put, with only the y-vectors changing direction, which gives the operation a total of 3 (6 x- and z-vectors stay put [+6] while 3 y-vectors flip [–3] for a total of 3). The other mirror plane ($\sigma_v'(yz)$) causes the hydrogen atoms to swap, so again you look only at oxygen with an unchanged x- and z-vector [+2] and a changed y-vector [–1] for a total of 1. Assembling these, we get a total reducible representation (Γ_{tot}) for water as follows:

E	C_2	$\sigma_v(xz)$	$\sigma_v'(yz)$	
9	-1	3	1	$= \Gamma_{tot}$

Remember, a reducible representation simply means it's composed of irreducible representations; for water, every irreducible representation has a value of 1 under E. This means the reducible representation with a 9 under E must be composed of nine irreducible representations; also, using $3N$ with N = three atoms, water has 9 degrees of freedom. Coincidence? Of course not! The reducible representation always renders the sum of all the molecule's modes of motion (translational, rotational, and vibrational).

Looking at the contribution per atom

You may have noticed that both mirror planes contribute 1 for every unshifted atom; $\sigma_v(xz)$ had three unshifted atoms while $\sigma_v'(yz)$ had one unshifted atom. The contribution per unshifted atom is constant regardless of the number of atoms involved. This fact comes in handy when you have molecules with dozens of atoms. Instead of keeping track of three vectors for every atom, you can simply focus on how many atoms stay put when you perform a given symmetry operation. You just saw that σ is always 1, E is always 3 (because nothing changes), and i is always –3 (because everything changes). C_n and S_n are a little more tricky because the amount of rotation affects the contribution per atom. You can memorize the formulas $C_n = 2\cos(360°/n) + 1$ and $S_n = 2\cos(360°/n) – 1$, or you can just memorize the C_ns and S_ns you will regularly encounter, as shown in Table 7-1.

Table 7-1	The Contribution per Atom for Various Molecular Symmetry Operations to Determine Unshifted Atoms								
Symmetry operation	E	σ	i	C_2	C_3	C_4	C_6	S_3	S_4	S_6
Contribution per atom	3	1	–3	–1	0	1	2	–2	–1	0

Using this technique, you can quickly calculate Γ_{tot} for water by examining simply the number of atoms that stay put. For E and $\sigma_v(xz)$, all three atoms are unshifted; for C_2 and $\sigma_v'(yz)$, the hydrogen atoms swapped so only oxygen goes unshifted. Now you simply multiply the number of unshifted atoms by the contribution of atoms to determine Γ_{tot}:

E	C_2	$\sigma_v(xz)$	$\sigma_v'(yz)$	
3	1	3	1	= Unshifted atoms
3	-1	1	1	= Contribution per atom
9	-1	3	1	= Γ_{tot}

If you're like me, you may hate memorizing anything (I still check the bands of my underwear to remember my name at times). Instead of memorizing the stuff in Table 7-1, you can simply sum up the irreducible reps that have the linear functions x, y, and z associated with them. For water, B_1 has x, B_2 has

y, and A_1 has z. If you add up the sum of these three irreducible representations, you will get 3 –1 1 1, exactly what you would expect your contribution per atom to be. If you rely on this method, make sure that you always get 3 for E, or you have made a mistake (usually on something with degeneracy). For example, on a D_{4h} character table, you will find the linear combined function (x, y) for E_u (which is doubly degenerate). Because that irreducible representation includes both x and y, you would add that only once with A_{2u} (the irreducible representation with z in it).

Pulling out the irreducible reps

Now that you have Γ_{tot}, you must determine the irreducible representations that make up each of your degrees of freedom. To do this, you use the following equation:

$N_i = (1/h)*$ the sum of (the coefficient of the class)(the character of the reducible representation)(the character of the irreducible representation) for every class

where N_i is the number of irreducible representations for any given irreducible representation (i) and h is the order of the character table. For water, you have four irreducible representations, so you would need to find how many times each appears in Γ_{tot}:

$$N_{A1}: (\tfrac{1}{4})[(1)(9)(1) + (1)(-1)(1) + (1)(3)(1) + (1)(1)(1)] = 3\,A_1$$
$$N_{A2}: (\tfrac{1}{4})[(1)(9)(1) + (1)(-1)(1) + (1)(3)(-1) + (1)(1)(-1)] = 1\,A_2$$
$$N_{B1}: (\tfrac{1}{4})[(1)(9)(1) + (1)(-1)(1) + (1)(3)(1) + (1)(1)(-1)] = 3\,B_1$$
$$N_{B2}: (\tfrac{1}{4})[(1)(9)(1) + (1)(-1)(-1) + (1)(3)(-1) + (1)(1)(1)] = 2\,B_2$$

So the Γ_{tot} for the water molecule is composed of nine irreducible representations (as you expected), specifically: $3\,A_1 + 1\,A_2 + 3\,B_1 + 2\,B_2$

Determining translational modes

To determine the translational modes, simply see what modes have a linear function of x, y, or z. For water, these functions lie in B_1, B_2, and A_1, respectively. Subtracting these from the total irreducible representations, you now have $A_1 + B_1 + B_2$ for your three translational modes with $2\,A_1 + 1\,A_2 + 2\,B_1 + 1\,B_2$ remaining for rotational and vibrational modes.

Focusing on rotational and vibrational modes

Similar to the translational modes, you simply want to determine what modes have a linear function of R_x, R_y, or R_z. For water, these are associated with B_2, B_1, and A_2, respectively. Subtracting these from the remaining modes, you now have $A_2 + B_1 + B_2$ for your three rotational modes with $2\,A_1 + 1\,B_1$ remaining for vibrational modes.

Infrared and Raman active modes

Two valuable instruments in the chemist's arsenal to identify and study molecules are infrared (IR) and Raman spectrometers. In IR spectroscopy, you hit your sample with a spectrum of infrared radiation covering many frequencies (IR is essentially just light with lower frequency than visible light). If any of the frequencies match the frequency that the atoms are vibrating, the bond actually absorbs that frequency and the detector opposite of the IR source notices that not all the frequencies came through. Whereas IR measures changes in the IR you pump through your sample, the Raman measures changes in visible light (usually from a laser) that gets scattered (changes directions) when hitting your sample. For this reason, you move the detector from opposite the light source as in IR to a 90° angle to the sample.

As a quick rule, if a vibrational mode has a linear function of x, y, or z associated with its irreducible representation, it is *IR active* (the IR spectrometer can see it). If it does not have one of these functions, then the mode can't be detected by IR. Similarly, *Raman active* vibrational modes exist only if the irreducible representation has a quadratic function (xy, xz, yz, x^2, y^2, z^2) or a combination of these terms (such as $x^2 + y^2 + z^2$).

Looking back at water, your three vibrational modes ($2 A_1 + 1 B_1$) are all IR active (A_1 lists z and B_1 lists x) and are all Raman active (A_1 lists x^2, y^2, z^2 and B_1 lists xz).

Chapter 8

Ionic and Metallic Bonding

*B*oth ionic and metallic bonding are similar to each other due to the efficient packing of ions in solid materials. They do this by trying to maximize the number of nearest neighbors with which each ion is in contact. In ionic compounds the ions share electrons with their nearest neighbors, but in metallic materials the ions share electrons in a delocalized manner.

In either ionic or metallic solids the ions pack together as much as possible. You can see examples when you are at the supermarket, look at how oranges and apples stack together to take up as little space as possible. Or as another example, take the case of sugar cubes in a box. Notice how cubes fit neatly into the box when they are packed evenly one on top of the other. This is how the ions try to pack together in both ionic and metallic materials. And this is what we talk about in this chapter (about ions, not sugar cubes).

Blame It on Electrostatic Attraction: Forming Ionic Bonds

The difference between covalent crystals, such as diamond (see Chapter 6), and ionic crystals, such as table salt, is due to how neighboring atoms share electrons: Covalent crystals share electrons in overlapping orbitals, whereas atoms forming ionic crystals donate electrons to the neighboring atom. Ionic bonding is the complete transfer of an electron from one atom to another.

The ionic model treats the material like it's made up of oppositely charged spheres that interact by Coulomb forces. These forces deal with electrical charges and the distance between each charge. Ions are treated as spherical point charges that are either positively or negatively charged. There are possibilities for attracting ions together (opposite-charged ions attract each other), and there are possibilities for repelling ions apart (similar-type charges repel each other).

Electronegativity is a measure of how much an atom in a molecule attracts electrons, compared to an electropositive atom that's willing to donate electrons instead. When one atom is electropositive, another is electronegative, and they are mixed together, there's a complete transfer of one or more electrons from one ion to the other (of opposite charge), and the solid compound that forms is regarded as *ionic*. The atoms do this to achieve the electron configuration of a noble gas to neutralize the electronic charge around the atom. The ions go from being ionic (charged) to being atoms (uncharged).

Electropositive ions are known as *cations*, and electronegative ions are called *anions*. As a general rule of thumb, for most ionic compounds, the anions are larger than the cations. This becomes important later in this chapter when we talk about the kinds of structures that can be made.

There are a few simple items to remember throughout this chapter, and they can be summarized in the following three points:

- The *ionic bond* involves the bonding of oppositely charged ions by complete donation of electrons from an electropositive atom to an electronegative atom.
- A *cation* has a positive charge.
- An *anion* has a negative charge.

In ionic compounds, the greatest stability is achieved by packing the most amount of anions around each cation, and vice versa.

Marrying a cation and an anion

Any marriage requires some give and take. In the case of ionic bonding, it's the give and take of electrons. The classic example of ionic bonding involves table salt, or sodium chloride (NaCl), which is made up of sodium cations and chlorine anions.

The Na atom has one electron in the outer valence shell and the Cl atom has one vacancy in the outer valence shell. The ions are attracted to one another because of their opposite charges. Figure 8-1 illustrates the relationship between a cation and an anion in an ionic compound.

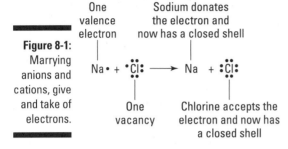

Figure 8-1:
Marrying
anions and
cations, give
and take of
electrons.

One valence electron

Sodium donates the electron and now has a closed shell

One vacancy

Chlorine accepts the electron and now has a closed shell

Atoms don't just give up or take on electrons freely, there are energy considerations to take into account. Generally speaking if there is a large energy penalty for cation formation that isn't recovered from taking on the negative charges, then an ionic bond will not form. Due to this, ionic bonds often form with metals because they have low ionization energies. In practice many ionic compounds have both covalent and ionic bonding.

Measuring bond strength: Lattice energy

Chemists calculate the *lattice energy* of an ionic compound to characterize the strength of the bond of the solid material. The lattice energy is the energy required to separate the solid ionic compound into gas-phase ions. By applying Hess's law and following the steps of the Born-Haber cycle, you can estimate the lattice energy of an ionic compound. Hess's law states that the amount of heat needed to change one substance to another depends on the substances and not on the reactions involved.

Lattice energy is based on the assumption that ions are acting as distinct and spherical entities, like billiard balls of different size (r) with some sort of electronic charge for each one (z). (See Chapter 18 for more on lattice energy.) They have non-directional bonding between them, which includes both the electrostatic attractive and repulsive forces. This is based on the way they are treated according to Coulomb forces.

Remember how we said that the ions act like spherical balls that try to pack together as neatly as possible? That they have as many nearest neighbors as possible? Well, in this picture, now you can also add the fact that the bond angle between the ion and cation doesn't make a difference in the strength of the bond.

When these ions form into solid materials, they're described according to how they pack together with respect to the positions that the spheres have relative to a cube. The cube that defines the simplest and most basic unit is called the lattice, and the energy it takes to make atoms/ions form into that lattice is known as the lattice energy.

Considering table salt again, the lattice energy, U_L, is the energy needed to transform NaCl back to Na and Cl as gases. This equation describes the reaction:

$$NaCl\ (s) \rightarrow Na^+\ (g) + Cl^-\ (g)$$

Strictly speaking, the quantities involved are the *enthalpy* rather than energy, so it makes more sense to write H_L for *lattice enthalpy*. This changes the equation to:

$$H_L(NaCl) = -\Delta H_f(NaCl) + \Delta H_{at}(Na) + \tfrac{1}{2}B(Cl_2) + I(Na) - A(Cl)$$

This equation states that the lattice enthalpy of sodium chloride ($H_L(NaCl)$) is equal to the negative enthalpy of formation of NaCl ($-\Delta H_f(NaCl)$), plus the enthalpy of atomization of Na solid($\Delta H_{at}(Na)$) plus half the bond enthalpy of Cl_2($\tfrac{1}{2}B(Cl_2)$), plus the ionization energy of Na ($I(Na)$) minus the electron affinity of Cl ($A(Cl)$).

$I(Na)$ is greater than $A(Cl)$ in the equation, which shows that Na and Cl atoms in the gas phase are more stable than the ions Na^+ and Cl^-. The lattice energy stabilizes the ionic charge distribution in solid NaCl. A similar result is found for all ionic solids. The lattice energy can be calculated quite simply when you know the values of the ionic radii and charge on the atoms that are involved. The following equation illustrates that the ionic radii, and the ionic charge, are important factors to be considered in the calculation:

$$H_L = A\left(\frac{z^+z^-}{r^+ + r^-}\right) - B$$

Where A is the binding constant, z^+ is the charge on the cation, z^- is the charge on the anion, r^+ is the cation radius, r^- is the anion radius, and B is the constant of repulsion.

As you move from left to right across the periodic table, the atoms increase in ionic charge (z value); and as you move down the periodic table, the ionic radii (r value) decreases. Thus, the value of the denominator ($r^+ + r^-$) in the equation increases, meaning that the total value increases going from left to right. The higher the total value, the more likely a reaction is to occur. As the z value rises, the numerator gets larger, and it gets larger much faster because it is the product of both z values, as opposed to the denominator that's the sum of the radii values. As a result, the elements situated further from the left side of the periodic table are more likely to form ionic crystals because they typically have greater lattice energies. This explains why there are so many halides formed, and also it explains why the stability of Group 2 halides is greater than the Group 1 halides. This is because the Group 2 halides have a much higher lattice energy because the z values for Group 2 atoms are larger than those of Group 1 atoms.

Enthalpy versus energy

Energy is an important concept in science, especially in chemistry. There are many forms of energy, and there are many consequences for a system when energy is added or removed. In chemistry it's important to know about the energy that affects chemical reactions, and for this reason we use enthalpy to help guide us. The word *enthalpy* comes from Greek for "inner warmth," and it has the symbol H. When enthalpy changes and ΔH is a positive value, the process is *endothermic* (it absorbs heat). But when the value for ΔH is a negative value, then the process is *exothermic* (heat is released). Enthalpy deals with the energy of a reaction that takes place under constant pressure (isobaric), but variable temperature and volume. This type of situation resembles what happens with gaseous reactions, so it's used in the Born-Haber calculation to determine the energy that is required, or released, when the atoms/ions change from a gas to a solid (desublimation). (Chapter 18 has more on the Born-Haber calculation.)

In the case of NaCl, the calculation is successful and matches well with experimental data. For example, the lattice energy of NaCl has been found to be −787 kJ/mol, and it has been calculated to be −772 kJ/mol.

Coexisting with covalent bonds

Ionic solids often contain ions that display some degree of covalent bonding as well as ionic bonding, especially because some ionic materials can exhibit polarity in their electron orbitals.

When an electron orbital undergoes *polarization*, the electron cloud can get distorted in such a way that the electrons become concentrated on one side of the atom. This kind of distortion deforms the orbital and makes it sharper at one end, thus giving the orbital some directionality. When this occurs, a covalent bond is likely to form. Figure 8-2 illustrates the difference between an ideal ionic bonding pair (with equally polarized orbitals) and a bonding pair with polarization great enough to result in a covalent bond instead of an ionic bond.

More covalent compounds (such as carbonates) are less soluble because the lattice energy is much higher, so they don't dissociate readily and become soluble.

In a typical covalent bond, there's electron sharing and directionality. If there is less electron sharing and more electron donating, then the bond has an ionic-covalent character. To quantify the ionic-covalent bonding character, chemists ask what would happen if there were to be some amount of charge displacement occurring.

Figure 8-2:
The effect of polarization on an idealized ionic pair (a) with no polarization, compared to when both ions are equally polarized (b), and the case where the polarization is great enough to form a covalent bond (c).

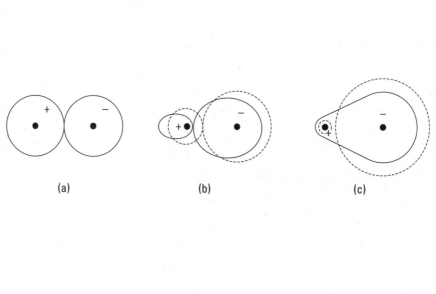

(a)　　　　　　(b)　　　　　　(c)

Start by looking at the bonding triangle in Figure 8-3. In this triangle, the three types of bonding (ionic, covalent, and metallic) are represented. For compounds with a simple AB formula, you can work out what kind of bonding is most likely based on the respective Pauling electronegativity. In real atoms, there is a combination of ionic and covalent character. The degree of the ionic bonding, or the percentage ionic character, can be determined for a particular system. If the difference in the electronegativity between two atoms in a solid is high (3 to 4), the system is 80 to 100 percent ionic; but if the difference in electronegativity is low (0 to 1), the system is 0 to 20 percent ionic.

In Figure 8-3, follow the dashed lines to see which bonding is most likely with a compound such as aluminum trichloride ($AlCl_3$) when aluminum is reacted with chlorine.

In reality, the bonding between atoms can be based on a combination from all three bond types.

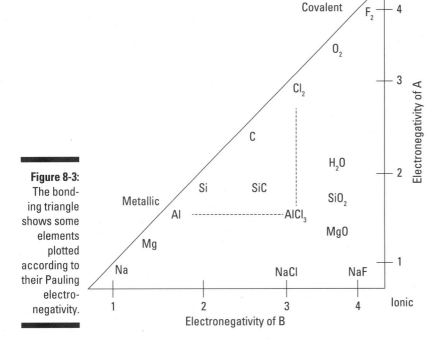

Conducting electricity in solution

Electrical conductivity in a material occurs because electrons are able to flow. If electrons can't flow through a material, it is not conductive. There's a close relationship between conductivity and chemical bonding in solids. Each type of bond creates distinctive electrical properties. The best conductors have electrons that move freely, because they are only loosely bound in the compound.

When an ionic compound is made, there is a one-time electron transfer, and after that the electron becomes strongly bound so that no conductivity is possible. For example, a lump of dry table salt won't conduct electricity. But if you dissolve the salt compound in water, the separated ions float away from the bulk crystal while in solution. Now they are able to help water conduct electricity, because they are charged and can respond to an electric field. Normally, water doesn't conduct electricity, but with ions, or electrolytes present, the solution conducts electricity. This is especially useful in the brain where electrical signals from nerve impulses are used to make the brain function, without an adequate amount and the right types of electrolytes the brain doesn't function properly, because the electrical signal does not get transmitted correctly.

Unless ionic compounds are dissolved into a solution, they share electrons in such a way that doesn't conduct electricity. Metallic bonds, on the other hand, do conduct electricity, even in solid state. See the section "What Is a Metal, Anyway?" for more details on metal bonding.

Admiring Ionic Crystals

Ions can exist in solution as well as in a crystalline, or solid form. Ideal ionic crystals are composed of atoms which may be represented as hard spheres, of varying size and opposing charge. Ionic crystals are often brittle because electrons are taken up by anions, and they often have high melting points.

The formation of ionic crystals is calculated using the Born-Haber cycle. It's called a *cycle* because ultimately the compound can cycle back and forth between its crystal form and its free ion form. When you go on a road trip, you want to be able to go home when you're done traveling. Just as the atoms can break apart and become soluble in an appropriate solvent, they can recombine (aggregate) to form a solid again under the appropriate conditions. Lattice energy plays a major role in the formation of ionic crystals.

An energy diagram (enthalpy) called the Born-Haber cycle is used like a road map to show all the possible side reactions that can occur; like with any road trip, there are multiple ways to get to a destination. Flooding on the road, for example, can cause you to take a different route; this is analogous to the way the environment affects the reaction dynamics.

The types of structures that form can't be predicted by the Born-Haber calculation. Instead, you have to familiarize yourself with the various lattice systems that are possible. The difference in the lattice system makes a big difference toward the ultimate appearance and properties of the material.

Studying shapes: Lattice types

When you break down a crystal you find that it fits into a certain mold, and that mold is known as the unit cell of a particular crystal system. Ionic materials form into one of seven different lattice types, or crystal systems that can be used to describe every ionic crystal.

We describe each crystal system and include *crystallographic notation* to help distinguish the characteristics of each type of crystal system. Figure 8-4 shows a diagram with a box that represents the basic unit of a crystal system. The box sits on a Cartesian axis and has six sides of lengths a, b, and c. The angles between each side are given by:

α = the angle between b and c

β = the angle between a and c

γ = the angle between a and b

a, b, c, α, β, γ are collectively known as the *lattice parameters* — often also called *unit cell parameters*, or just *cell parameters*. (Visit Chapter 18 for details on the unit cell and its role in solid materials.)

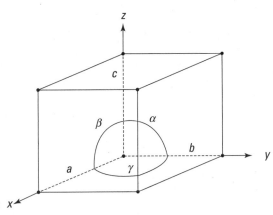

Figure 8-4:
Directional vectors of an ionic crystal: a, b, and c with the angles between them denoted as α, β, and γ.

There are seven crystal systems which ionic crystals have been found to form. They represent the way in which the bond angles and the bond lengths can differ in crystals. The cystal systems can be seen in Figure 8-5. Here are the details of the seven crystal systems:

- ✔ **Cubic:** Most metals, solid hydrogen, and table salt adopt the cubic system. This is the simple box model, all the lengths of the sides are equal, and the angle between each side is equal, too; therefore $a = b = c$ and $\alpha = \beta = \gamma = 90°$.

- ✔ **Hexagonal:** Magnesium, zinc, quartz, and solid oxygen adopt the hexagonal system. In this case, the atoms spread out in such a way that the structure looks like a hexagon because it has six sides parallel to one another. In this case, $a = b \neq c$ and $\alpha = \beta = 90°$, $\gamma = 120°$.

- ✔ **Tetragonal:** Sodium sulfide and indium adopt the tetragonal system. This is similar to the cubic structure, but it's stretched along one of the axes, so now not all of the sides are of equal length; therefore, $a = b \neq c$ and $\alpha = \beta = \gamma = 90°$.

- ✔ **Trigonal:** Quartz can adopt this system. The trigonal structure is also like the simple box analogy, but imagine that instead of the box standing up straight you have pushed it so that it leans to one side all the time. In this case $a = b = c$ and $\alpha = \beta = \gamma \neq 90°$.

- ✔ **Orthorhombic:** Sulfur forms orthorhombic crystals. This is like a stretched box, but with one of the sides being wider than the other ones, sort of like the shape box that a flat screen TV is delivered in. In this case, $a \neq b \neq c$ and $\alpha = \beta = \gamma = 90°$.

- ✔ **Monoclinic:** Household sugar takes the monoclinic form. This is like the orthorhombic structure, but instead of standing up straight it's as if the box is tilted (like the trigonal structure). In this case, the dimensions are $a \neq b \neq c$ and $\alpha = \gamma = 90°$, $\beta \neq 90°$.

- ✔ **Triclinic:** Boric acid takes the triclinic system. The triclinic structure is one of the more complicated structures. Because none of the lengths or any of the angles are the same, the dimensions are $a \neq b \neq c$ and $\alpha \neq \beta \neq \gamma \neq 90°$.

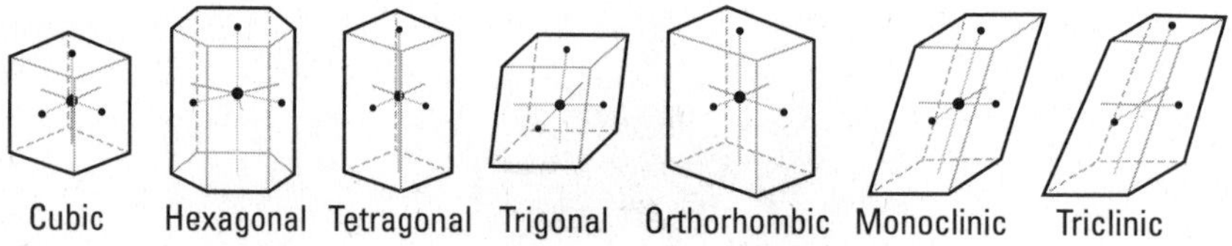

Figure 8-5:
The seven
crystal
systems.

Cubic   Hexagonal   Tetragonal   Trigonal   Orthorhombic   Monoclinic   Triclinic

Size matters (when it's ionic)

Earlier in this chapter, we said that generally speaking anions are larger than cations. In this part of the chapter, we look at why the size matters. In short, this is because the solid structure is most stable when each ion has as many nearest neighbors as possible. The number of nearest neighbors can be affected by the size difference between the anion and cation.

As a general rule of thumb, the coordination number determines how many neighbors the ions can bond with, but in practice this is not always the case. For example, when the size difference is very large between the anion and cation. Let's say that the anion is so big that only two anions can physically fit around the cation. Even if it has a coordination number of 4, it can only bond with two ions.

As you can see it really depends on the size difference between the anion and cation. Another way of saying this is that it depends on the ratio of sizes between each of them. There's a radius ratio rule that can be used to help determine what kind of structure is created. Table 8-1 shows where r_c and r_a represent the radius of the cation and anion, respectively.

Table 8-1	Radius Ratio Rule and the Arrangement of Ions	
r_c/r_a	*Arrangement*	*Number of Nearest Neighbors*
1.000	fcc or hcp	12
0.732-1.000	Cubic	8
0.414-0.732	Octahedral	6
0.225-0.414	Tetrahedral	4
0.155-0.225	Trigonal	3
0.155	Linear	2

fcc and hcp represent face-centered cubic, and hexagonal close-packed structures, respectively. (These are demonstrated in Chapter 18.)

The radius ratio rule can only serve as a guide, however; it's based on the assumption that the ions act as hard spheres, and that there is no overlap between the spheres. Strictly, this isn't correct, as we have mentioned previously. There is some amount of covalent bonding that can occur, and this is where the radius ratio rule starts to break down. But it serves as a good tool to get an idea of what the structures will be like.

"I'm Melting!" Dissolving Ionic Compounds with Water: Solubility

There are a large number of factors that influence the *solubility* of atoms. When a solid material is soluble, it is able to be dissolved into a liquid (often, but not always, water). Lattice energy, ion size, and hydration energy play an important role in the solubility of ionic materials.

If the material is composed of ions of very different sizes, it is readily soluble in water; at the same time, a compound of similar sized ions is less soluble in water. For a compound made up of MX it is soluble in water when the radius of M is smaller than X by 0.8 Å.

The *Gibbs energy* plays a large role in the solubility of materials also (see Chapter 16 for details on Gibbs energy). In the following reaction:

$$MX \rightarrow M^+ (aq) + X^- (aq)$$

the M and X must overcome the lattice energy that's holding the solid together and break it apart to be soluble. The energy to do this can come from ion hydration and other entropic forces.

Just add water: Hydrated ions

When a metal or ion is dissolved in water, the atoms find themselves in the center of an ever-increasing ring of water molecules that keep the ion hydrated. The water molecule is *polar*, having a negative and positive side so they orientate themselves around the ions to balance their charges.

For example, if the ion is positively charged like the case of many metals, then the oxygen part of the H_2O molecule points toward the metal center, whereas the hydrogen part points away from the metal center (see Figure 8-6). Several H_2O molecules bunch around the metal to *passivate*, or neutralize, the charge, forming a sphere around the ions.

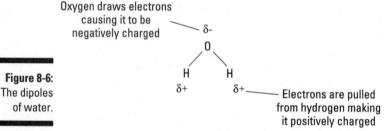

Oxygen draws electrons causing it to be negatively charged — δ-

Electrons are pulled from hydrogen making it positively charged

Figure 8-6:
The dipoles of water.

REMEMBER

The number of water molecules around the central ion is known as the *hydration number.* There are many other water molecules in the vicinity also affected by the presence of the positive charge from the metal, so there are several spheres of water molecules circling the metal center. The circle around the metal is actually a three-dimensional sphere of atoms around the metal center, and it's called a *hydration sphere.* Figure 8-7 illustrates a metal center surrounded by two hydration spheres.

Ions that are very large with a low-charge density have fewer hydration spheres. And ions that are small and have a high-charge density may have many hydration spheres around the central ion.

The *hydration energy* is the energy needed to form a hydrated ion. The difference in *hydration energy* between the cation and anion pair also affects the solubility. When the difference between the two is negligible, or close to zero, the materials are soluble in water. Figure 8-8 shows a series of ionic and metallic compounds along with their hydration energy and degree of solubility.

Figure 8-7:
Hydration
spheres
created by
metal center
surrounded
by water
molecules.
The dashed
line marks
the border
between
the first
and second
hydration
sphere.

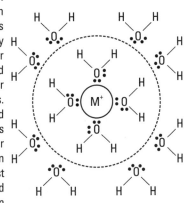

Figure 8-8:
The solubil-
ity of various
compounds
compared
to the dif-
ference in
hydration
energy for
each of
the atoms
present.

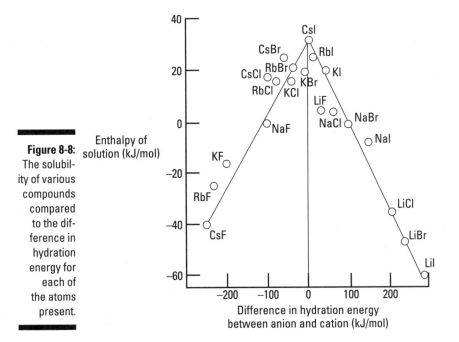

Counting soluble compounds

While studying or practicing inorganic chemistry, it may be helpful to keep in mind these eight rules of solubility:

✔ All alkali metals are soluble; they come from Group I.

✔ All salts with ammonium (NH_4^+) are soluble.

✔ All nitrates (NO_3^-), chlorates (ClO_3^-), perchlorates (ClO_4^-), and acetates ($C_2H_3O_2^-$) are soluble.

✔ All chlorides (Cl^-), bromides (Br^-), and iodides (I^-) are soluble with the exception of Ag^+, Pb^{2+}, and Hg_2^{2+}.

✔ All sulfate (SO_4^-) compounds are soluble, except that Ba^{2+}, Sr^{2+}, Ca^{2+}, Pb^{2+}, Hg_2^{2+} Hg^{2+}, Ca^{2+}, and Ag^+ are only slightly soluble.

✔ All hydroxide (OH^-) compounds are insoluble unless they are of the alkali metals, or made with Ba^{2+}, Ca^{2+}, and Sr^{2+}.

✔ All sulfide ($S^=$) compounds are insoluble, except for those that are made with either the alkali or alkaline earth metals.

✔ All sulfites (SO_2^-), carbonates ($CO_3^=$), chromates (CrO_4^-), and phosphates (PO_4^{2-}) are insoluble, with the exceptions of those that are made with NH_4^+, or the alkali metals.

What Is a Metal, Anyway?

Most of the periodic table is made of metals. The science of metals is called *metallurgy*. More recently, this has come to be known as *solid state chemistry* and *solid state physics*, because it includes materials not classically considered metals, but act like metals, such as conductive polymers that are known as synthetic metals for example.

Most metal compounds are solid at room temperature, with mercury (Hg) being the only liquid metal. Metals have high melting and boiling points because the atoms are tightly packed in the crystal. (This also results in metals having high density because of the tight packing.)

Metals have only a few valence electrons in the outer shell, usually no more than 4. These are given up during a chemical reaction to form metallic, positive ions. Metals form structures that have many nearest neighbors. They are solids with high coordination numbers and in which the highest occupied energy band (conduction band) is only partially filled with electrons. The electrical conductivity of metals generally decreases with temperature. For more information about the conduction band, you can read up on molecular orbital theory in Chapter 6.

Tracing the history of metallurgy

Metals have been known to humankind since antiquity. The first use of metals goes as far back as 6000 BC.

Gold (ca): 6000 BC

Copper (ca): 4200 BC

Silver (ca): 4000 BC

Lead (ca): 3500 BC

Tin (ca): 1750 BC

Iron, smelted (ca): 1500 BC

Mercury (ca): 750 BC

Gold was the first to be discovered; ever since then, it hasn't ceased to amaze mankind. So much so that it's written into our language to define something that is universal, such as the "golden rule" or a "gold standard."

Admiring the properties of solid metals

A *solid metal* is defined as having many delocalized electrons that can be shared by all the other metal atoms in the material. So when sodium forms a crystal, each atom comes into contact with eight other sodium atoms. It's outermost electron is in the $3s^1$ orbital, and is delocalized throughout the crystal. Magnesium has two electrons in the 3s orbital that both participate in delocalization, which lowers the energy potential of the electrons around the orbital. This means that magnesium has a higher melting point than sodium and is also a lot stronger. This is due to the fact that the delocalized electrons add extra stability to the material because they participate in the bonding of the metal atoms all together.

The way that the metals are bonded and what metals are present can cause the materials to have a wide variety of properties. Metals are known for their strength, which is why they're used in heavy machinery. But they can also be malleable and thus can be made in to certain shapes without damaging the crystal or making the structure weaker. They're also ductile in the way they can be stretched, especially when they are heated. They are usually opaque and have some kind of color that we can see, but they can also reflect back and absorb other types of radiation and not just the electromagnetic radiation that we see in the visible spectrum. Some of the light that's not absorbed can be reflected, and this is what gives them a nice luster.

Metals are unique because they can be both extremely strong and durable, yet also very easy to work with and fashion into specific shapes. This makes them very useful for a lot of applications, from machinery to tools to jewelry.

Strength

The strength of a material is determined at the atomic scale based on how much stress a material can undergo along a certain axis with respect to the direction of the force being exerted on it, before it breaks or cracks. Materials can be bent sideways, squeezed along different directions, and undergo sheer forces also.

The strength of metals comes from the fact that the atoms are all neatly packed together in a uniform manner. Strength can be increased by forming alloys, or mixtures of different metals.

Malleability

Malleability is the measure of how a material can be bent and shaped in various ways without breaking or coming apart. Usually this means that the material is somewhat weak.

Metals are very malleable, which is why they can be beaten into different shapes. This is the opposite of ionic crystals, which are brittle, and break easily.

Ductility

Metals are *ductile*, which means they can be drawn out into a wire. Metals are ductile because of the presence of delocalized electrons. These electrons can move about quite freely because they aren't tied to one particular metal atom. When the metal is pulled apart or smashed, the electrons aren't forced to line up against each other; if they did, they would repel each other, and the material would be brittle. This is what happens in ionic materials because the electrons are more tightly bound and can't move away from each other when they are brought closer to one another.

Opacity

This is the property that describes how electromagnetic radiation interacts with the material. Materials that are opaque usually don't let light pass through them, and can also block other forms of electromagnetic radiation. All metals have some degree of opacity; many of them would stand out in an x-ray picture because they strongly interact with the x-ray beam, which in effect stops the x-rays from passing any further. The opacity arises because of the high density of the atoms.

Luster

This is a property that describes a material's capacity to reflect light, this gives gold and silver lots of shine. This is often the wow factor of these precious metals, as this is the property that makes it shiny and gives the reflected light some added brilliance.

Delocalizing electrons: Conductivity

Perhaps the most well-known, useful, and fascinating property of metals is their ability to conduct electricity and heat. This ability is the result of delocalized electrons that often reside on the surface of the materials. The electrons are able to move around quite freely, and in the presence of an electric field they can also be made to flow in a certain direction. These flowing electrons can carry an electrical charge, so the material is *conductive*.

When two electropositive atoms come together they form a metallic solid. At the atomic scale, the metallic compound is held together by metallic bonds, which are relatively weak. This allows the electrons to move around freely. This characteristic of metals is called the *delocalization of electrons.*

The delocalization creates a sea of electrons that can move freely between the metal atoms and therefore allows electrical current to travel through it or along it.

Delocalization of electrons occurs when electrons are shared between electropositive metal atoms. There is an uneven number of electrons present so the electron is delocalized and can *conduct*, or carry current.

There are three mechanisms for carrying current across metals:

- **Scattering:** Random motions of electrons bumping into each other; this causes heating to occur in metal and leads to the breakdown of cables when operating at very high voltages.

- **Ballistic:** In a ballistic conductor the mean free path of a conduction electron is quite long, so it can travel a large distance without undergoing scattering events that would otherwise slow it down.

- **Hopping:** The electron bounces on the surface from one site of delocalized electrons to another. In energy terms, the electron moves via jumping across potential barriers from site to site along the surface of a material.

Analyzing alloys

Metals are so useful because they can be fashioned into particular shapes that fit a particular function. Consider the difference between the shape of a copper pot to that of a piece of copper wire. The properties that make them so useful are the ductility, malleability, and the strength. But when you have solid materials with mixed metal compositions, the properties of the material can be greatly enhanced. This process is known as *alloying*, and in short it's the science of marrying metals together into solid materials so that they have specific properties that make them desirable for particular tasks.

Finding it with Eureka!

One day several thousand years ago, a bright scientist by the name of Archimedes had to resolve an interesting problem for his king, because his king was suspicious that a crown was not made from pure gold, but was tainted with some other metal.

The king, Hiero, suspected that a goldsmith had added silver to a gold crown instead of making it from the pure gold the he was supplied with. The goldsmith was given a specific mass of gold to use in the crown, but the king was suspicious some of it was missing. Hiero asked Archimedes to work out if the crown was made of pure gold, or if it was made with a mixture of the two. The crown was considered holy, so the king did not want Archimedes to break it because he didn't want to offend the gods.

Archimedes devised a rather simple but cunning method to test the crown. Archimedes noticed that when he got into his bath tub at home, the level of the water would rise and fall according to when he got in or out, to such an extent that the water could even overflow from the bath. Upon this discovery, he apparently ran down the street naked screaming "Eureka," which meant "I found it." What he had done was notice how a dense material displaces a certain volume of water. When two objects of equal mass are placed in the water, the less dense material displaces more water, because it has a larger volume. Knowing that the density of gold and silver are different, gold is more dense than silver, he was able to determine that some silver was used in the making of the crown. He placed a mass of pure gold that weighed the same as the crown into a bucket, and then filled it to the brim with water. When he took the gold out, the water level fell accordingly. So when he put the crown into the water, it caused the water to overflow. Even though it had the same mass as the pure gold, it had a larger volume, as evidenced by the fact that the water only overflowed when the crown was in it. He reasoned that the crown must be tainted with silver because it had a larger volume compared to the pure gold.

Alloys are materials that are formed with two or more metals bonded together. Alloys form due to the similarity between the metals. The word alloy itself stems from Latin *alligare*, which means to bind. This was adapted in 16th-century French in the word *allier*, which means to combine.

Certain alloys are preferable because they can have better resistance to corrosion, they are stronger, and they can have other beneficial properties besides. It's also quite common to find a precious metal alloyed with a less precious metal. This could be done by a jeweler because the more precious metal is more expensive, but when mixed with a similar-looking metal, only experts can tell that they have been mixed.

Alloys are most often made using the transition metals such as iron, or chromium for example. Since they form hard and strong materials with high melting points and high boiling points, they are often used for machinery and other structures that require strength and endurance. Commonly used household alloys include both bronze (copper and tin) and brass (copper and zinc). The transition metals can form alloys with one another and with other nonmetallic atoms also, for example when carbon is added to iron in small amounts ($\leq 2.1\%$) it forms steel. But when small amounts of chromium are added to iron at about 10.5 percent of the total mass, then you form stainless steel.

When liquid metals are combined, there is some limit to how much the metals are soluble or miscible in each other. If they're mixed in concentrations below that limit they blend perfectly together to form an alloy. For example, an alloy of copper and zinc can be made when the zinc concentration is less than 40 percent of the total content, and the resulting alloy has Cu and Zn atoms uniformly distributed around each other. Generally speaking, alloys are often stronger than the metals on their own because the second metal restricts the motion of the atoms, preventing them from sliding thus making the material stronger.

The formation of an alloy follows three simple rules that are known as the Hume-Rothery rules.

- ✔ The atomic radii of both metals must be within 15 percent of each other; this prevents extra strain on the lattice.
- ✔ The crystal structures for both of the metals should be the same so that they bond in the same crystal arrangement.
- ✔ You need identical valences and similar electronegativity of the metals.

The strength of the alloy can be determined by the size of the atomic radius of the second metal, compared to the size of the atom for the host metal. The greater the difference, the greater the strength of the alloy. In the case of bronze, a tin atom has a radius of 154 picometer (pm), and copper a size of 128 pm. But if zinc is added to copper so as to make brass, zinc has a size of 133 pm. Brass is weaker than bronze because the difference between copper and zinc (5 pm) is less than that of copper and tin (26 pm).

Swimming in the Electron Sea: Metallic Bonding Theories

Metallic bonding is quite different from other bonding motifs. There is no ionic contribution, and it's impossible to have covalent bonds with adjacent pairs of neighboring atoms. There are not enough electrons, and as such, there are not enough orbitals to bond with. In the macroscale, the bonding is understood according to band theory. But remember that the band theory is the combination of all the molecular orbitals (Chapter 6) of all the metals that are bonding with each other. Metal-metal bonding can be with one other metal only. In the case that there are just two metals bonded to each other the MO theory shows that metal-metal bonding can be done by σ, π, δ orbitals.

Looking at a transition metal, such as iron which has both an s-orbital overlap and a d-orbital overlap, notice that the more electrons that overlap, the tougher or stronger the metal is. Also, not only do the atoms cling to each other strongly through electron overlap, but they also fit together very tightly. As such, the crystal packing constants of metals are in close-packed cubic lattices such as face-centered or body-centered structures. Having this kind of close packing gives the material greater strength.

Another characteristic that makes metals different from ionic bonded materials is that in a solid metal all the atoms are typically of the same element. For example, they are all made of copper or iron. Being of one element alone, or made up of elements that have similar properties (such as alloys), gives the atoms the chance to slide around from lattice to lattice without breaking the crystal; however, an ionic crystal (table salt) that's made of two oppositely charged ions can't allow the atoms to move around as much due to coulombic repulsion. They are oppositely charged leading to situations with two negative ions beside each other (or positively charged ions), and they would repel each other. Remember, opposites attract, and like charges repel. The repulsion of the similarly charged ions could be enough to make the material break apart.

Similarly for covalent compounds, SiO_2 (glass) is held together by covalent bonding, it is strong, but under stress the atoms can't slide or move; so the glass shatters. In contrast, consider lead, composed of only lead atoms. You can hammer it, bend it, and it does not break easily because the lead atoms can move around without causing big changes in the crystal structure. In essence, glass is hard and brittle; lead is soft and malleable.

The three most prominent theories to date to explain metal bonding include free-electron theory, the valence bond theory, and band theory. They represent a progression in scientists' understanding of metals over time.

Free-electron theory

A theory developed in the 1900s that depicts metallic cations, surrounded by a sea of electrons. The oldest theory used, the free- electron theory explains a great deal about the properties of the materials, but is still only a qualitative assessment and doesn't offer any method for testing or quantifying the theory. Figure 8-9 illustrates the free-electron model of metals.

This model may account for various properties observed in metals such as:

✔ **Malleability and ductility:** The sea of electrons act like a sponge, so when the metal is hammered the composition of the metal structure is not harmed or changed. The positively charged metal nuclei may be rearranged, but the sea of electrons simply adjust to the new formation and keep the metal intact.

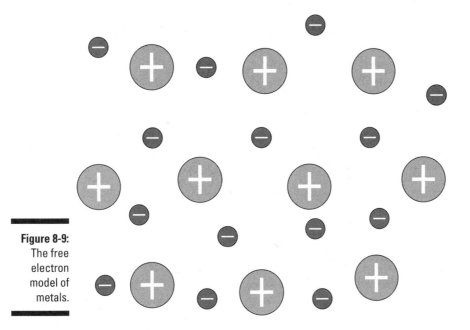

Figure 8-9:
The free electron model of metals.

> ✓ **Luster:** The free electrons in the sea can absorb photons making them opaque-looking. However, the surface electrons also reflect light back at same frequency of incidence making the surface shiny.
>
> ✓ **Conductivity:** Because the electrons are free, the electrons move through the wire, going in one end and out another end carrying electrical current.

Because this model can't be tested and is largely considered as oversimplified, it has fallen out of favor with most chemists. Continued research expanded on this theory focusing not just on the free electrons around the metal atom, but on singling out the electrons that are found in the valence bond only.

Valence bond theory

The valence bond theory suggests that the bonding in metals occurs due to the overlap of localized electrons only (see Chapter 6 for a more in-depth description of valence bond theory). Because the electrons are considered to be localized, they are essentially pinned down in space, making it somewhat easier to calculate the electron wave function. This theory was developed to explain the waveform, or more precisely, the wave function of atoms. Wherever the wave functions for atoms overlap, this represents bond formation. When it was applied using hydrogen as a model, two new concepts regarding chemical bonding were developed — resonance and orbital hybridization.

Resonance describes how a single bond and double bond that are close to one another in a molecule can switch places and go back and forth. This is a regular occurrence for many molecules, so the theory of resonance was important in explaining the behaviors of lots of materials.

Orbital hybridization describes what happens when the electron orbitals of two different atoms mix with one another. It's useful in explaining important chemical and physical properties of many materials such as the conductivity in synthetic metals and single-walled carbon nanotubes, for example.

Although the valance bond theory can't be used to fully describe metals and was later replaced using molecular orbital theory (Chapter 6), its development was an important step in chemistry research nonetheless.

Band theory

Today, the generally accepted concept that is used to explain metal bonding comes from an extension of molecular orbital (MO) theory (see Chapter 6 for more details). When you take two atoms and bond them together you get two molecular orbitals, one bonding and one anti-bonding. Then if you take three atoms together, you get three MOs because you have another MO that lies between the other two. Each new set of electrons that are added to the molecules causes a new MO to open up, eventually with enough atoms bonded together so that all the molecular orbitals overlap each other. The overlap causes them to form a *band*, and this can be seen in Figure 8-10.

Figure 8-10: When many molecular orbitals overlap in a large system, they form a band of electrons.

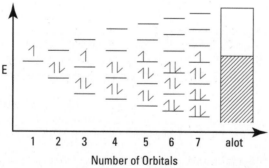

This happens in crystals where many atoms are bonded. There can exist a gap in the band that represents the energy difference between electrons in the valence band with respect to those of the conduction band. The size of the band gap can have important implications for use in optics and electronics. For more about band gaps, see Chapter 18.

Chapter 9

Clinging to Complex Ions: Coordination Complexes

. . . .

In This Chapter

▶ Understanding what makes a coordination complex

▶ Realizing how isomers are the same and yet different

▶ Following the rules to name coordination compounds

▶ Dissolving compounds in water: Addition salts

▶ Looking at coordination complexes in different element groups

▶ Applying coordination chemistry to the real world

. . . .

Coordination complexes, also called coordination compounds, describe the various types of complex molecules built around a metal atom center. This means, of course, that coordination chemistry studies all the metals and metalloids that exist in the periodic table. This covers nearly 90 percent of all the elements. Fortunately, it's not necessary to look at each element individually when studying coordination complexes because certain trends exist that serve as a guide, regardless of which elements are involved.

In this chapter, we explain how the coordination number of the central ion functions to guide the chemistry of coordination compounds, and how molecules are shaped when they have that specific coordination number. We also describe how some complexes built of the same elements form different shapes, or isomers, and how properties such as color have played an important role in the field of coordination chemistry. Last but not least, we explain how metal complexes are classified and how chemists name coordination compounds.

Metal complexes and *coordination compounds* are synonymous with each other, and the terms are used interchangeably for that reason. At the heart of every coordination compound is a metal atom, surrounded by electron donors called *ligands*. These ligands, bound directly to a metal center are considered to be within the *inner coordination sphere*.

The coordination number is used to count the number of atoms or ions that are bonded to a central metal atom/ion. For short, it's written as CN and can range from 2 up to 12, depending on the size of the central atom/ion relative to the atoms/ions or molecules that are bonded to it. For example if an atom has a coordination number of 2, then it can bond with two ligands. A small metal atom can pack those atoms/ions much tighter than a larger metal center, but if it had a CN of 6, for example, it may not be able to fit all six ligands because it's too small. The consequence of this is that even if metals have the same CN, it's common to find that metals with larger atomic radii often exhibit greater coordination numbers in practice. Transition metals can form complexes with coordination numbers from 2 to 9. If the metal ion is a lanthanide or actinide, it can get even higher coordination numbers. (See Chapter 14 about these special elements.)

Counting bonds

The complex ions that form the basis of coordination compounds accept electrons from surrounding atoms or ions to form coordinate covalent bonds (see Chapter 6). A ligand must have at least one pair of electrons to donate to form such a bond, but in some cases may be able to donate multiple pairs of electrons (especially if the ligand is a multidentate ligand — electron pairs may be donated from multiple atoms at the same time). Chemists use the following names to distinguish among ligands:

- **Monodentate:** Ligands that donate only one pair of electrons to form a bond with a metal atom (mono means one, and dentate means bite).

- **Multidentate:** Multidentate describes ligands that donate two or more pairs of electrons to the metal atom (each pair is donated from a different atom in the ligand molecule).

- **Chelates:** Complex ions with multidentate ligands are considered a special class called *chelates*. Due to the multiple bonds formed by chelating complexes, the atoms of the *chelating agent*, or multidentate ligand, wrap around the metal and hold onto it, somewhat like holding a juggling ball between your finger and your thumb.

Seeking stability

Chemists use the *stability constant* to characterize the strength and stability of coordination complexes. The stability constant depends on two factors, and they are thermodynamic and kinetic stability.

Thermodynamic stability refers to the energy difference in a reaction between the reactants and the products. This approach can be used to calculate the equilibrium constant, and this can give a good idea about what products are most likely to form.

For simplicity, let's take an example of a metal ion, M, that's bonded with two monodentate ligands (l). This metal ligand complex is now Mll (has two ligands, therefore two l's). Now imagine we react this complex with a bidentate ligand, that's identified as capital L for now. This results in the following reaction:

$$Mll + L = ML + l + l$$

Almost invariably you find that the monodentate ligands are replaced by the bidentate ligand. This is known as the *chelate effect*, and it arises because the stability of the complex is greater with the bidentate ligand.

This can be calculated using numbers, but the working principle is based on entropy. Remember, this is about thermodynamics after all, and thermodynamics deals with entropy, or disorder, within a system. You see, what happens is that when the bidentate ligand binds to the metal and displaces the two monodentate ligands, the entropy, or disorder, changes. For example, look at the fact that before the bidentate ligand was added to the metal there were just two species involved (Mll and L). But when the bidentate ligand binds to the metal, the disorder in the system increases, because you now have two ligands that are free and one complex; in total this is three components (ML, l, and l); yes, two of these are the same. In this case you go from a situation with two species, to a situation with three species, and this is how the chelate effect works.

There is also kinetic stability, and this deals with the reactivity of the compounds, and for the most part this refers to the speed at which a ligand can bind to a metal atom, and the speed for a ligand to be released from the metal atom. This is called a ligand exchange. In short, it describes the situation where one ligand is replaced by another.

Water is a good ligand and has been used to determine water exchange rates for many metals, and metals with a variety of oxidation states. The values can serve as a general guide to gauge what you might expect to happen for a certain metal ion as per the oxidation state. It's particularly useful if you want to compare the relative reaction rates of two different metals, or to determine the relative reaction rates of the same metal but with different oxidation states.

For the most part, but not always, as the metal oxidation state increases, the rate of exchange decreases. Water exchange is also referred to as *water replacement* in some textbooks.

Transition metals and d-block elements form stable complexes that are more stable than the other elements. Metal ions require a combination of three factors to form a stable complex, all of which are common in transition metals and d-block elements:

- ✔ A high ionic charge (+2, +3, and higher) so that donated electrons from the ligand are attracted by the positive nucleus of the cation.

- ✔ A small ionic radius so that the ligands can get closer for maximum binding affinity and as a result, stronger bond strength.

✔ Partially filled external, or frontier, orbitals are available for bonding with the ligands. The presence of the free orbitals allows for the formation of a covalent complex. This is because the electrons that are donated by the ligands are placed in the vacant orbitals, and this is why the d-block atoms form covalent bonds with ligands. And this is why the transition metals form the majority of organometallic complexes that are known. This is important later when talking about d-orbital splitting and crystal field theory.

Grouping geometries

The shape of a coordination complex is a result of the bonding arrangement of the ligands around the metal center of the complex ion. The way that the ligands are attached to the metal atom causes the entire compound to adopt a certain shape (see Figure 9-1). The shapes are associated with the coordination number and the size of the ligand. There are a few common shapes:

✔ **Linear:** Linear coordination structures are formed by ligand-metal-ligand, or L-M-L, bonding in a line. They are usually associated with CN of 2 and +1 cations, such as Cu, Ag, Au, and Hg_2. Sometimes they occur for other transition metals when the ligands are extremely bulky, leaving room for only two ligands. Examples include BeH_2, CO_2, and HCN.

✔ **Square planar:** Square planar structures form with CN of 4, and all the ligands are in the same plane as the center metal. Examples include BrF_4^-, ClF_4^-, and XeF_4.

✔ **Tetrahedral:** Tetrahedral structures are also associated with CN of 4 and are more common than square planar structure. In tetrahedral or *tetragonal* structures the four ligands and central atom are not in the same plane. The tetragonal shape can form for all the nontransition metals, and some transition metals. Examples include CH_4, CCl_4, and SO_4^{2-}. (The T-metals on the right side of the d-block have no electron in the valence shell and can only form sigma bonds to the ligands. For more about σ (sigma) and π (pi) bonds, see Chapters 6 and 7.)

✔ **Octahedral:** Octahedral structures usually form with CN of 6. These complexes form an octahedron, with near perfect symmetry that can be elongated if a variety of ligands are used. This is important because nearly all cations form 6-coordinate complexes, so it's the most common geometric shape. Examples include SF_6, and SiF_6^{3-}.

✔ **Trigonal:** Trigonal structures form with CN 3 and CN 5. These complexes can form planar trigonal structures, or trigonal pyramidal shapes. These are less common because coordination numbers of 3 and 5 are rare. Examples include BF_3, Co_3^{2-}, and $CoCl_2$.

Figure 9-1:
Ball and stick diagrams of common coordination geometries.

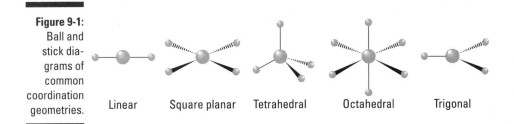

Linear   Square planar   Tetrahedral   Octahedral   Trigonal

Some structures exhibit a trait called *fluxionality*. Trigonal bipyramid and square pyramid complexes associated with CN 5 as well as pentagonal bipyramid, capped octahedron, and capped trigonal prism associated with CN 7 are flexible. Compounds with these geometries can switch between shapes easily. This characteristic is called *fluxionality* because the shape of the structures are in flux, and are able to change.

Compounds with higher coordination numbers, although rare, do occur and form more complex geometric structures. The highest coordination numbers, however, don't conform to any regular geometry.

Identifying Isomers

Coordination compounds often form isomers. *Isomers* are compounds with the same formula (or components) but different physical structures. There are two broad classes of isomers — constitutional, or structural isomers, and stereoisomers. Within these two categories, there are specific types of isomers, with different degrees of complexity. Figure 9-2 illustrates with a flow-chart how the different types of isomers are related to one another.

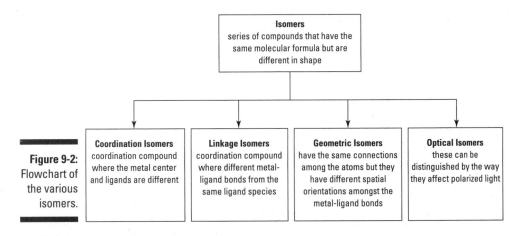

Figure 9-2:
Flowchart of the various isomers.

Isomers
series of compounds that have the same molecular formula but are different in shape

Coordination Isomers
coordination compound where the metal center and ligands are different

Linkage Isomers
coordination compound where different metal-ligand bonds from the same ligand species

Geometric Isomers
have the same connections among the atoms but they have different spatial orientations amongst the metal-ligand bonds

Optical Isomers
these can be distinguished by the way they affect polarized light

Connecting differently: Structural isomers

Constitutional or structural isomers display a difference in *connectivity*. This means that although they have all of the same atoms, the atoms are connected to one another in different ways. Within the broad category of structural isomers there are two more specific classes of isomers:

- ✔ **Coordination isomers:** Coordination isomers have a different distribution of ligands between metal centers. These commonly occur where both the anion and the cation contain a metal, and either of them can act as a coordination center. Another way coordination isomers form is when a ligand from the complex ion switches place with the counterion during bonding. Other examples include when you have the same metal, but different ligands present, or the same metal but in different oxidation states.

- ✔ **Linkage isomers:** Linkage isomers occur when a ligand species can attach to the metal atom in different orientations. For example, when thiocyanate (SCN^-) bonds to a metal, in one isomer the metal can be linked through the nitrogen atom whereas in the other isomer it's linked though the sulfur atom. (The orientation of these links affects the name of the coordination compound, which we describe in the next section.)

Arranged differently: Stereoisomers

The other broad class of isomers is stereoisomers. Within this class of isomers, the atoms are connected in the same way, but have different spatial arrangements. Within this class, chemists split them further into geometric and optical isomers.

Geometric isomers

Geometric isomers have the same connections between the atoms, but they have different spatial orientations among the metal-ligand bonds. For example, with a square planar molecule such as $Pt(NH_3)Cl_2$, the ligands can either be situated across from each other (180 degrees apart) or next to each other (90 degrees apart).

When the ligands are beside each other, the compound is a *cis* isomer. When the ligands are across from one another, the compound is a *trans* isomer. Cis isomers have the ligands occupying adjacent corners of the molecule; trans isomers have the ligands right across from each other in the same molecule. The names stem from Latin — *cis* means *next to*, and *trans* means *across*.

Propellers are chiral

Did you know that propellers are chiral? Take the example of a single inboard engine on a boat that has a three-bladed propeller. When you put the engine into reverse see which way the boat starts to turn. For example, if you have a left-handed propeller, it pulls the stern to port; in other words, it pulls the front of the boat to the left (as you face forward). But if you have a right-handed propeller, it pulls the stern to starboard; this is pulling the front of the boat to the right (if you are facing forward). Port and starboard are nautical terms that are used to signify the left- and right- hand side of a vessel,

they are used instead of saying left and right, because left and right depend on which way you are facing. Whereas port and starboard don't depend on which way you are facing.

A propeller is chiral, it just pushes water in different directions. In chemical terms, when a molecule is left-handed it's denoted as Λ (capital lambda), and when it's right-handed it's denoted as Δ (capital delta). In the following figure, you can see how a three-bladed propeller is either left-handed (on top), or right-handed (on the bottom).

Left-handed

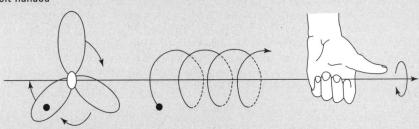

Right-handed

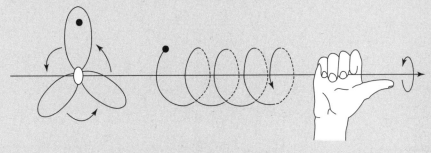

Cis and trans molecules are different compounds with different properties even though they have the same chemical formula. For example, cis-$Pt(NH_3)Cl_2$ is a polar molecule and is more soluble in water compared to trans-$Pt(NH_3)Cl_2$. This is because in the trans isomer the Pt-Cl and the Pt-NH_3 point in opposite directions, the charges cancel each other out, making it non-polar. These two isomers are illustrated in Figure 9-3.

Figure 9-3:
Cis-platin
and trans-
platin.

cis platin   trans platin

Optical isomers

Optical isomers are compounds that are not superimposable on their own mirror image. This can be a little tricky to conceptualize if you are new to the idea. Take a look at your hands and you can see that they are mirror images of each other, one is a reflection of the other. This characteristic is called *chirality*, and objects that display it are described as *chiral*. Now take one hand and place it on top of the other and try to match or superimpose it. You see that they don't match. Because of their chirality, your hands aren't super-imposable on each other.

Optical isomers display this same feature because they are chiral. Molecules and objects that do this are called *enantiomers*. An enantiomer is a molecule whose shape or structure has mirror images that aren't identical to each other. If the molecule isn't chiral, it's referred to as *achiral*. This concept is illustrated in Figure 9-4.

Figure 9-4:
Enantiomers
and handed-
ness.

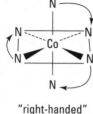
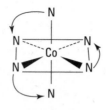

"right-handed"   "left-handed"

Enantiomer compounds have identical chemical reactivities except for their reactions with other chiral substances and how they affect plane polarized light. Enantiomers cause polarized light to rotate. If the light is rotated to the right, it gets the (+) symbol; if it is rotated to the left, it gets the (−) symbol.

Polarizing light

The polarity of light is different from the polarity of a chemical. Light waves work by spreading out from a light source in such a way that the electromagnetic vibrations are occurring in all planes parallel to the direction of the light. When the light becomes polarized, the vibrations are no longer occurring in all planes, but rather get narrowed down to just one plane. This minimizes the amount of light, therefore also lowers the intensity of the light. This is used in certain pairs of sunglasses, for example. They can be made to minimize the sheen from the surface of water because light that's reflected from a horizontal or flat surface becomes horizontally polarized. The lenses of these sunglasses have a layer of photosensitive molecules that are aligned preferentially in one direction that only allow vertically polarized light to pass through. In short, the arrangement of the molecules is designed to cancel out the light that is reflected from flat surfaces such as water, this makes it easier to see while fishing or boating.

This is useful when determining which enantiomer you are working with in a laboratory. Simply shine plane polarized light through a solution to see which way the light is then rotated. Because different enantiomers shift the light one way or the other, this can tell you which ones are present. There is, however, another possibility that can occur; there may be equal amounts of each enantiomer present and in this case, the effect on the light would cancel out. This is known as having a *racemic* mixture, and it produces no net change in the rotation of the light.

Naming Coordination Complexes

To get a good census for all the coordination compounds, you must be able to name them according to the chemical formula of the complex. The naming of coordination compounds is now largely understood according to the rules set out by the International Union of Pure and Applied Chemistry (IUPAC). Prior to this standardization, the naming scheme was based on the color of the compound, but that was later found to be unreliable. The following are the seven steps to naming coordination compounds.

1. Cations are named before anions (similar to ionic materials).

2. The ligands are named in alphabetical order, after that the name of metal is listed. Anionic ligands have the ending –o, for example F^- is fluoro, Cl^- is chloro, and Br^- is bromo. Positive ligands have the ending –ium, such as NH_2-NH_2 is hydrazinium. Neutral ligands have no special naming scheme, for example NH_3 is amine, H_2O is aqua, and CO is carbonyl. And for covalent ligands you can employ their common names, such as phenyl, methyl, or cyclopentadienyl.

3. The ligands are numbered next. If there are two or more of the same kind of ligand present, they are counted together and added up. The number of times can be from 2 to 12 and is indicated by using a numerical prefix. Here are the first few to get you started: di-, tri-, tetra-, penta-, hexa-. But if the ligand itself has one of these prefixes in the name already, such as the ligand *di*pyridyl, you should instead use bis-, tris-, tetrakis-, or pentakis-; these numerical prefixes should also be used for polydentate ligands.

4. The naming of the central metal atom depends on the charge. If it's a negatively charged metal, the anion is given the suffix –ate, such as colbaltate, zincate, sulphate. But if the metal is either neutral or positively charged, it's just named as the element.

5. The oxidation state of the metal atom has to be shown; it's written as a Roman numeral inside parentheses: (IV), for example.

6. There's the possibility of having more than one metal atom present. These are called polynuclear atoms, and the ligand bridging the metal atoms is indicated by the prefix –μ.

7. Ligands can attach through different atoms. Take the examples of $M\text{-}NO_2$, nitro, and M-ONO, nitrito. The metal is bonded to the nitrogen in the first case, and it's bonded through an oxygen in the second case. Even though both have the same chemical formula, they're arranged differently. This is what happens with linkage isomers such as thiocyanate (SCN^-). When bonded according to M-SCN, it's called thiocyanato; but when it's like M-NCS, it's called isothiocyanto.

The following are a few examples of coordination compounds:

$K_3[Fe(CN)_6]$: Potassium is the cation, so it's named first. There are six cyanide ligands that are each anionic so it's referred to as hexacyano. The iron atom is negatively charged, so it has the ending –ate, and it has a +3 oxidation state so 3 is written in Roman numerals in brackets such as (III). In total this compound is called potassium hexacyanoferrate(III).

$Fe_4[Fe(Cn)_6]_3$: Iron ferrocyanide, more commonly known as Prussian Blue, is a famous compound with a brilliant greenish-blue color. It inspired chemists to understand the mechanisms by which atoms produce color. To artists, it serves as a pigment and still today is readily available. It's also known as Chinese blue and milori blue.

$Fe(C_5H_5)_2$ **Bis(cyclopentadienyl)iron(II):** This is more commonly known as iron pentacarbonyl, and it's a famous sandwich complex. A sandwich complex is similar to a filling (metal atom) that's sandwiched between two flat molecules; the concept is itself worth a Nobel Prize (see Chapter 21).

$Cr_2O_3 \cdot 2H_2O$: Chromium hydroxide hydrate, known as viridian, has a green color and can be used as a pigment in paint.

You also need to become familiar with the common type of ligands that you will encounter, because they're often written as abbreviations instead of being fully spelled out. Table 9-1 shows some of the more common ligands.

Table 9-1		Common Ligands		
Type	_Electron Charge_	_Ligand Name_	_Formula of Ligand_	_What They're Called_
Monodentate	Neutral	Ammonia	NH_3	Ammine
		Water	H_2O	Aqua
		Carbon Monoxide	CO	Carbonyl
		Pyridine	pyr	Pyridine
	Minus One	Azide	N_3^-	Azido
		Bromide	Br^-	Bromido
		Chloride	Cl^-	Chlorido
		Cyanide	CN^-	Cyanido
		Fluoride	F^-	Fluorido
		Hydride	H^-	Hydride
		Hydroxide	OH^-	Hydroxido
		Alkoxide	OR^-	Alkoxide
		Nitrite	NO_2^-	Nitrito
		Thiocyanate	SCN^- or NCS^-	Thiocyanato
Bidentate	Neutral	Bipyridine	bipy	Bipyridine
		Ethylenediamine	en	Ethylenediamine
	Minus Two	Carbonate	CO_3^{2-}	Carbonato
		Oxide	O^{2-}	Oxo
		Oxalate	$C_2O_4^{2-}$	Oxolato
		Sulfate	SO_4^{2-}	Sulfato

Some chemists categorize coordination compounds in more general terms, according to the ligands that are used. For example:

- ✔ **Werner complexes:** Werner complexes are the classical examples of coordination compounds. The ligands bind to the metals almost exclusively through lone pairs of electrons on the donating atom of the ligand. Common ligands include H_2O, HN_3, Cl^-, CN^-, and ethylenediamine (en, for short).

- ✔ **Organometallic complexes:** Organometallic complexes use organic ligands, such as cyclopentane, alkenes, and others. Sometimes the ligands are *organic-like*, such as phosphines, hydrides, and carbonyls (CO). Turn to Chapter 15 for more details on organometallic compounds.

- ✔ **Bioinorganic complexes:** Bioinorganic complexes include ligands that form in nature. These include side chains of amino acids, and other ligands that bond to proteins. Turn to Chapter 17 for more details on bioinorganic chemistry.

- ✔ **Cluster complexes:** Cluster complexes have ligands that fall into the previously listed categories but are unique in that they can also have a second metal that can be treated as part of the ligand system.

Sorting Out the Salts

There are two distinct types of coordination compounds separated from one another by their reactivity, which is due to the nature of the bonding from the metal to the ligand. *Complex compounds* are bound by coordinate covalent bonds described by valence bond theory (see Chapter 6). *Addition salts* or *double salts*, however, are bound according to electrostatic interactions, or ionic bonds. (Turn to Chapter 8 for details on ionic bonds and salts.)

A *double salt* compound is stable in a solid state, but *dissociates*, or dissolves, into ions in an aqueous environment. They are called double salts because they are usually composed of two metal elements. (See Chapter 8 for details on the solubility of salts.) Complex compounds are also stable in solid state, but when added to an aqueous solution they maintain their structure rather than dissolving into ions. The addition salts or double salts also typically have lower coordination numbers than the complex compounds.

Understanding the difference between the two types of coordination compounds is useful for various chemistry applications. The addition salts can be used to separate other materials from a solution. When the ions of a salt are dissolved into solution, they can bind to another ligand in the solution, forming an insoluble compound that precipitates out of solution.

For example, the compound $[Ln(NO_3)_3][3Mg(NO_3)_2] \cdot 24H_2O$ is a double salt that's used in the *fractional crystallization* (removing it from solution by forming insoluble compounds) of different lanthanide elements (the lanthanide elements are covered in Chapter 14). This occurs because there's a difference in the bonding between the different lanthanide elements in how strongly they bind to the magnesium ion, so when the ion is in solution, it attracts certain lanthanide elements more strongly than others. The new lanthanide-magnesium ion compounds are insoluble and precipitate from the solution.

Creating Metal Complexes throughout the Periodic Table

As we described in Chapter 2 the periodic table is useful because it organizes the elements into groups that share similar characteristics and reactivities. As a result, it's useful to look at the different element groups to think about how they form coordination compounds with different characteristics and reactivities. In this section, we describe the characteristics of the coordination complex in terms of the different element groups across the periodic table.

Alkali metals

Metal complexes with alkali metals can form according to both covalent, and non-covalent interactions between the metal center and the ligands. For these s-block elements it's common to find that the compounds are made with the metal having oxidation numbers that are lower than the metal group number. They form cluster compounds (see Chapter 15) such as Rb_6O, Cs_4O_4, for example.

Lithium is also frequently used as alkyl lithium compounds. These are useful in substitution reactions for organic chemistry and have the unique capacity to invert the polarity of the functional carbon atom. Upon reduction, it's converted from an electrophilic species to a nucleophilic species.

Alkali earth metals

With the alkali earth metals s-block elements, the bonding is typically ionic in nature. However Mg^+ can have covalent type bonding, and beryllium atoms bond with some polarization.

Ring around the ions

The development of cryptands and crown ethers is a relatively new endeavor in the history of chemistry, and has led to the field of *supramolecular chemistry*. Cryptands and crown ethers have been developed with the intention of selectively chelating with the s-block elements. Some of these are radioactive, and this is one way they can be removed from water supplies. They are called *crown ethers* because they look like a spiky crown with a central space that can be filled with a metal. An *ether* is an organic molecule that has an oxygen center with hydrocarbon attached to it on either side. The crown ethers are closed-loop versions, and their names are based on both the number of atoms in the ring, and the number of oxygen atoms present. One example is 18-crown-6. Another ring compound, hydroxypropyl beta-cyclodextrin, is used in commonly found household odor removing cleaning products. It's a donut-shaped compound that traps odor molecules preventing their ready detection.

For these s-block elements, the stability of complexes is often affected by the difference in the size of the ionic radii relative to the cation and anion pair. Large anions are generally better stabilized by large cations; for example, barium peroxide is more stable than beryllium peroxide. In fact, the barium peroxide compound is so stable that it forms spontaneously in air.

There are several important covalent metal-alkali compounds that serve as catalysts for reactions between carbon atoms. One example is *Grignard reagents*, made from the bonding between lithium and alkyl groups. These are useful for metathesis reactions, whereby alkyl groups can be linked together. Organic chemists use this to form C-C bonds.

Transition metals

As we have described throughout most of this chapter, the bond between a transition metal and ligand is a covalent coordinate bond, where the number of bonds equals the coordination number of the complex.

Transition metal complexes can form with coordination numbers ranging from 2 to 9. They display such a diverse and interesting chemistry with a large range of possible compounds. We cover more detail on these compounds in Chapter 15 that deals with organometallic compounds.

The reactivity of the transition metal complexes changes as you go down the Group of the d-block elements. The ionic radius increases due to the fact that the electron cloud around the nucleus gets larger. This leads to weaker complexes because the ligands experience a greater repulsion from the electron cloud, and so they require strong *nucleophiles* to act as ligands. As you move across the series from the left to the right, the number of free orbitals decreases and so the ones on the far right side will form complexes less easily, and also often require strong nucleophiles to act as ligands.

For transition metals, the more common features of coordination complexes is summarized by the following: they work by using 3d orbitals, the most common oxidation numbers are 4 and 6, so they form tetrahedral and octahedral structures most often.

Lanthanides and actinides

The coordination numbers for the lanthanide and actinide element groups can be very large, much larger than that found for other metals.

The coordination compounds formed with lanthanide ions are generalized according to the fact that they operate using the 4f electron orbitals instead of d-orbitals. They commonly have coordination numbers of 6, 7, 8, and 9. The shapes that these complexes make are often trigonal prisms, square antiprisms, or dodecahedrons. The interaction between the metal center and the ligand is usually weak, and there is little preference for the direction of the bonding between the metal center and the ligand. The bond strengths change between the metal center and various ligands according to the change in the electronegativity of the ligands, such that the ligand strengths are highest for $Cl^->NO_3^->H_2O>OH^->F^-$. In solution, the complexes can undergo very rapid transformations, and the ligands can be exchanged very fast. This is mainly because there's not a strong covalent bond between the metal center and the ligands.

The actinide complexes are still being investigated because many of the elements don't last long enough for fruitful chemical reactions to occur; many of them are made synthetically and are highly radioactive. (See Chapter 14 for more details about the lanthanide and actinide elements.) Actinides don't have the chemical uniformity that is found across the lanthanide elements, even though they have a common oxidation of +3.The actinides have large atomic radii, and for this reason they also often have high coordination numbers because they have lots of room around the central atom to bond ligands.

Metalloids

The metalloid elements include boron, silicon, germanium, arsenic, antimony, tellurium, and astatine. Ligand exchange reactions complete much faster for silicon complexes when compared with carbon complexes because the silicon atom can form a hypervalent transition state that carbon cannot.

A hypervalent molecule is a stable compound made with a main group element from the third row, such as silicon, phosphorus, and sulfur. What's unique about hypervalent compounds is that they can have more than the eight valence electrons. For this reason, the halides of silicon are more labile, or flexible, than similar carbon halides. The most important compounds of silicon and germanium contain electronegative halogens, oxygen, and nitrogen. Silicon also forms organosilicon compounds, which are widely studied because they have commercial applications as water repellents, sealants, and lubricants. There are a number of metalloid carbides that are known for both boron and silicon; they're very hard, covalent solids.

Applying Coordination Complexes in the Real World

Many metals are used as chromophores. These can include, but are not limited to, cobalt, nickel, iron, chromium, tungsten, ruthenium, gold, copper, vanadium, titanium, manganese, zinc, aluminum, zirconium, palladium, and tin. Organometallic complexes with these metals are often used as coloring agents for a variety of materials. Coupled with the appropriate ligands, they can be made soluble in a slew of solvents and so can be distributed to wherever they are needed.

Solubility depends on using the appropriate ligand. The ligand can alter the solubility and, therefore, also the suitability for certain applications. There are many organometallic double salt complexes used in industry. In such a complex, there are two metal atoms surrounded by several ligands that are associated with the metals. The most common ligand is water. In aqueous solution, the ions are hydrated (see Chapter 8). Water is an example of a neutral ligand.

The formation of the complex can be considered according to an acid/base reaction where the metal (atom/ion) is the Lewis acid that is ready to accept electrons from the ligands, that are like Lewis bases. As for all acid base reactions, the formation of the complex can vary according to the conditions in which the reaction is being carried out. There's a balance constant that's used to describe the ability to form the complex; this is called the stability constant, and it's a measure of how strongly the metal-ligand complex is held together. Some coordination complexes bind so strongly that they're used as standards in analytical chemistry to determine the concentrations of certain metal analytes. One such compound that's often used is ethylenediaminetetraacetate (EDTA, for short), because it readily forms strong complexes with a number of metals, particularly Ca^{2+} and Mg^{2+} ions that are found in hard water.

Part III

It's Elemental: Dining at the Periodic Table

The 5th Wave   By Rich Tennant

"How we doin' over here? Anyone need their hydrogen atoms shaken up?"

In this part . . .

The development of the periodic table brought chemistry into the modern world. Prior to having an organized table of the elements, there were hundreds of theories about why atoms reacted in different ways. The periodic table erased this confusion and made it possible for anyone to understand the differences between elements, based on where they sit on the periodic table.

This part uses the element hydrogen as a guide to the concepts of the periodic table. This small and simple atom serves as the first step in your journey through the periodic table. We present all of the elements from the alkali and alkaline earth metals, to the main group elements, and even the span of transition metals that connect one side of the periodic table to the other. You also get a good look at the mysterious, rare, and exotic lanthanide and actinide elements.

Chapter 10

What the H? Hydrogen!

This chapter talks about hydrogen and how it sits in a unique place in the periodic table. Actually, hydrogen is quite hard to place in the table because it has some very unique features that we share with you. There are several isotopes of hydrogen, and they're all quite unique and can be used in many different ways. Hydrogen also makes interesting bonds that are known as (of all things . . .) hydrogen bonds, and this plays a role in far-reaching areas that span from energy storage to information storage in DNA molecules. All in all, hydrogen is one of the most abundant elements in the periodic table and has many uses in chemistry and industry.

Visiting Hydrogen at Home: Its Place in the Periodic Table

With an atomic number of 1, Hydrogen is the first element found on the periodic table. Though it shares a place with helium in the first period, the two elements are so different from one another that you rarely see them grouped together; even on the periodic table, you find them on opposite sides. Helium is inert, whereas hydrogen is extremely reactive. Helium is similar to all of the noble gases (Group 18), yet hydrogen shows some similarities to the alkali metals (Group 1), the carbon group (Group 14), and the halogens (Group 17), three groups that are very different from one another. For these reasons, coupled with the fact that it is the most abundant element in the universe, hydrogen has been awarded its own chapter. It has a very special story to tell.

Hydrogen is the simplest of all the elements, having only one proton and one electron. With only one electron, hydrogen has an electronic structure of $1s^1$. This electronic structure makes hydrogen similar to the alkali metals of Group 1 (Li is [He] $2s^1$, Na is [Ne] $3s^1$, K is [Ar] $4s^1$, and so on) with only one *s*-electron. It's important to note that while hydrogen sits atop Group 1 on most periodic tables, as shown in Figure 10-1, it's not considered to be an alkali metal here on Earth (only under the most extreme conditions will hydrogen behave like an alkali metal, such as the core of Jupiter and Saturn!).

Note: Hydrogen does behave like an alkali metal under extremely high pressure environments. The cores of Jupiter and Saturn are liquid metallic hydrogen.

PERIODIC TABLE OF THE ELEMENTS

1 H													1 H			1 H	2 He
3 Li	4 Be											5 B	6 C	7 N	8 O	9 F	10 Ne
11 Na	12 Mg											13 Al	14 Si	15 P	16 S	17 Cl	18 Ar
19 K	20 Ca	21 Sc	22 Ti	23 V	24 Cr	25 Mn	26 Fe	27 Co	28 Ni	29 Cu	30 Zn	31 Ga	32 Ge	33 As	34 Se	35 Br	36 Kr
37 Rb	38 Sr	39 Y	40 Zr	41 Nb	42 Mo	43 Tc	44 Ru	45 Rh	46 Pd	47 Ag	48 Cd	49 In	50 Sn	51 Sb	52 Te	53 I	54 Xe
55 Cs	56 Ba	57-71	72 Hf	73 Ta	74 W	75 Re	76 Os	77 Ir	78 Pt	79 Au	80 Hg	81 Tl	82 Pb	83 Bi	84 Po	85 At	86 Rn
87 Fr	88 Ra	89-103	104 Rf	105 Db	106 Sg	107 Bh	108 Hs	109 Mt	110 Ds	111 Rg	112 Cn	113 Uut	114 Fl	115 Uup	116 Lv	117 Uus	118 Uuo

57 La	58 Ce	59 Pr	60 Nd	61 Pm	62 Sm	63 Eu	64 Gd	65 Tb	66 Dy	67 Ho	68 Er	69 Tm	70 Yb	71 Lu
89 Ac	90 Th	91 Pa	92 U	93 Np	94 Pu	95 Am	96 Cm	97 Bk	98 Cf	99 Es	100 Fm	101 Md	102 No	103 Lr

Figure 10-1: Three locations that hydrogen may be found atop the periodic table: Group 1 (due to $1s^1$ configuration), Group 14 (due to half-filled orbital), and Group 17 (due to 1 electron shy of full orbital). Hydrogen is most commonly found atop Group 1 by convention.

Hydrogen's similarities to the carbon group (Group 14) arise because the elements of the carbon group have a half-filled shell of electrons. For example, hydrogen has one valence electron, only half the electrons needed for the electronic configuration of He; similarly, all of the carbon group elements have four valence electrons, exactly half of the eight valence electrons needed to be a noble gas. However, the majority of the chemical reactions in which hydrogen is involved resemble Group 1 or Group 17. Hydrogen's electronic structure resembles the halogens (Group 17), because all of the halogens are only one electron shy of a noble gas electronic structure.

For all of these similarities, hydrogen doesn't belong to any specific group in the periodic table. You often encounter hydrogen atop Group 1 on a printed version of the periodic table, but it could arguably sit above Group 1 or 17. Table 10-1 lists some of the reasons for and against placing it in either of those two groups.

Table 10-1 **Arguments For and Against Placing Hydrogen above Groups 1 and 17 on the Periodic Table**

Argument for Placement	*Argument against Placement*
Group 1 (Alkali Metals)	
Commonly forms positive ion (H^+)	Is not a metal
Has a single s electron	Does not react with water
Group 17 (Halogens)	
Is a nonmetal	Rarely forms negative ion (H^-)
Forms a diatomic molecule	Is extremely reactive

The history of hydrogen in a nutshell

The word *hydrogen* stems from the Greek words *hydro*, meaning water, and *genes*, meaning creator. First termed by Antoine Lavoisier in 1783, the name arose from the creation of water after burning hydrogen gas in air (a very violent reaction), repeating the studies of Henry Cavendish less than 20 years prior. This finding may seem straightforward, but at the time it undermined ideas stemming back to Aristotle that everything was composed of only four elements: earth, water, air, and fire. Ultimately, what Aristotle defined as matter corresponds to modern concepts of states and heat: earth represented solids, water represented liquids, air represented gases, and fire represented heat. Lavoisier's experiment of mixing air and fire to produce water truly defied this idea of classical elements. So as with so many important scientific discoveries, the exploration of hydrogen began, quite literally, with a bang.

Appreciating the Merits of Hydrogen

Hydrogen has countless properties that make it useful for countless applications. Instead of listing them all, here are the few of the properties that make hydrogen interesting to explore.

Available in abundance

There is more hydrogen than any other element in the universe, measuring in at about 92 percent of the matter of the universe. But in our atmosphere, the abundance is much less. It's found tied up in rocks and minerals on earth, and in the water of the oceans and seas. But it's only the tenth-most abundant element. Even though its elemental form is found in such small quantities on earth, it's extremely useful and versatile, as evidenced by the fact that it's found in more compounds than any other element.

Molecular properties

Hydrogen is the lightest gas known, existing as a colorless and odorless gas in a diatomic state (H_2). Because it's so light, it was the gas of choice used in giant airships resembling great balloons known as Zeppelins to cargo and carry people great distances. Unfortunately, hydrogen is also very flammable. For example, a common test for hydrogen in the lab environment is to simply place hot embers in the vial in which you expect to find hydrogen. If hydrogen is present, the addition of the embers gives a pop sound. However, this is only a tiny vial, not a massive Zeppelin. The reaction is very exothermic, and the results are most infamous due to the horrible Hindenburg disaster in 1937. The Hindenburg airships held over 200,000 m^3 of hydrogen gas to provide lift for the ship and its crew, passengers, and mail cargo. Although much debate exists as to the cause of the accident (theories range from engine failures to minor leaks to lightning strikes), what is certain is that the flammability of the hydrogen caused the event to be disastrous and resulted in the deaths of 36 individuals. Due to the accident, engineers were forced to move to a more expensive but nonflammable alternative, helium, to lift Zeppelins.

Hydrogen naturally exists as the diatomic molecule H_2 with a very strong covalent bond (bond energy of 435.9 kJ/mol). Under normal conditions, hydrogen isn't reactive because of the strong H-H bond. To split H_2 requires a large amount of energy; therefore, reactions requiring H are usually slow unless they are mediated by high temperatures or a catalyst such as from the transition metal series.

Nuclear spin

Spin is a fundamental property of elementary particles. Electrons, being an elementary particle, possess a spin that's important for magnetism. Similarly, protons (composed of elementary particles known as *quarks* and *gluons*) also possess spin. Hydrogen has one half spin due to the presence of only one proton in its nucleus. This property is important for the use of nuclear magnetic resonance (NMR) spectroscopy and magnetic resonance imaging (MRI). Both of these instruments rely on the concept of nuclear spin, because it is capable of absorbing and reemitting electromagnetic energy. NMR spectroscopy is used by chemists to determine numerous properties of chemicals, most commonly the chemical's structure. MRI, on the other hand, uses the same technology to image internal structures within the body. Being composed mostly of water (H_2O), hydrogen-based MRI is ideal for imaging the body because you have lots of hydrogen atoms to image.

Introducing Hydrogen Isotopes

Hydrogen comes in three isotopes: *protium* ($_1^1H$ or H), *deuterium* ($_1^2H$ or D), and *tritium* ($_1^3H$ or T). They all contain one proton, which, by definition, makes them hydrogen; then they possess either 0, 1, or 2 neutrons, respectively. Just as the cation of protium is called a *proton*, the cation of deuterium is called a *deuteron*, and the cation of tritium is called a *triton*. Collectively, a hydrogen ion of an unknown isotope is simply referred to as a *hydron*.

Protium is by far the most abundant at 99.986 percent. Deuterium is known as *heavy hydrogen* and has an abundance of 0.014 percent; tritium is a radioactive material with an abundance of 7×10^{-16} percent on earth; however, it's been proposed that larger amounts of tritium exist on the moon. Both deuterium and tritium undergo very exothermic fusion reactions, so they are used in both research and thermonuclear weapons. There's also the hopeful possibility that they might be used as a future energy source also.

All the isotopes have the same electronic configuration, so they exhibit identical chemistry; however, the reactions occur at different rates. For example, the adsorption of H_2 to surfaces is more rapid than that of D_2. And H_2 reacts 13 times faster with Cl_2 than D_2, because H_2 has a lower activation energy.

The difference in mass between the isotopes leads to various property differences; these differences are commonly referred to as *isotope effects*. Hydrogen is unique because the isotope effects are greater than for isotopes of any of the other elements. With only one proton, the addition of one neutron nearly doubles the mass. This underlies the large differences in the physical properties of hydrogen isotopes.

Investing in Hydrogen Bonds

Hydrogen can form both intramolecular bonds and intermolecular bonds. Intramolecular hydrogen bonds (X-H—Y) form between elements that are part of the same molecule. This can lead to certain molecules taking very specific shapes. Conversely, intermolecular hydrogen bonding occurs between two distinct molecules, such as the case between base pairs of the DNA molecule.

Hydrogen can form hydrogen bonds with electronegative atoms such as N, O, or F. The H-X—H hydrogen bond has an energy of 10-60 kJ/mol. This energy is quite weak when compared to a covalent bond, but is strong compared with intermolecular forces such as the van der Waals force. This is why it's so easy for water to flow, because the bonding between the neighboring water molecules is quite weak.

Forming a hydrogen ion

Remember that hydrogen has a $1s^1$ electronic configuration with only one electron around the nucleus, so when it loses that electron, it becomes H^+, and the particle that is left is the proton. H^+ forms the hydronium ion (H_3O^+) in water, as seen in Chapter 5.

The most common solvent for H^+ is water; this is where it's found in abundance on earth.

A hydrogen ion can be formed readily upon the reduction of silver and iodine:

$$Ag^+ + H \cdot \rightarrow Ag + H^+$$
$$2\,I^- + 2\,H \cdot \rightarrow 2\,H^+ + I_2$$

Creating hydrides

Binary hydrogen compounds, or *hydrides*, can be formed by three routes:

1. Direct combination of the elements: $2\,E + H \rightarrow 2\,EH$

2. Protonation of a Brønsted base: $E^- + H_2O \rightarrow EH + OH^-$

3. Metathesis of halide with a hydride: $E^+H^- + EX \rightarrow E^+X^- + EH$

The electronegativity of the second element determines the type of hydride, and also the type of bonding that occurs. Hydrogen has an electronegativity of 2.1, which is in the middle of the Pauling electronegative scale with francium at 0.7 and fluorine at 4.0, so it can form a variety of bond types. To simplify things, hydrides can be classified in three categories: ionic or salt-like hydrides, covalent or molecular hydrides, and metallic or interstitial hydrides. An interstitial compound can occur when a small atom, such as hydrogen, can fit in the empty space of a metal lattice. Put another way, imagine a metal being a huge stack of oranges at your grocery store stacked in an orderly pyramid. Now take a handful of green peas and allow them to sit in the empty spaces between the oranges. Those green peas would be interstitial peas.

With an electronegativity only slightly above the median value of all the chemical elements, hydrogen behaves like a weakly electronegative nonmetal. It forms ionic compounds with very electropositive metals and covalent compounds with nonmetals. It also forms metal hydrides with some of the transition metals. The three major hydrides it forms and with which elements are shown in Figure 10-2.

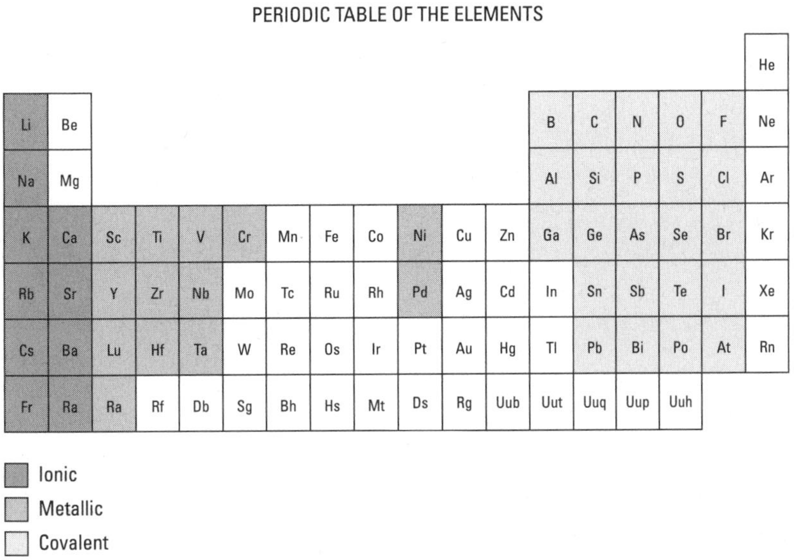

PERIODIC TABLE OF THE ELEMENTS

																	He
Li	Be											B	C	N	O	F	Ne
Na	Mg											Al	Si	P	S	Cl	Ar
K	Ca	Sc	Ti	V	Cr	Mn	Fe	Co	Ni	Cu	Zn	Ga	Ge	As	Se	Br	Kr
Rb	Sr	Y	Zr	Nb	Mo	Tc	Ru	Rh	Pd	Ag	Cd	In	Sn	Sb	Te	I	Xe
Cs	Ba	Lu	Hf	Ta	W	Re	Os	Ir	Pt	Au	Hg	Tl	Pb	Bi	Po	At	Rn
Fr	Ra	Ra	Rf	Db	Sg	Bh	Hs	Mt	Ds	Rg	Uub	Uut	Uuq	Uup	Uuh		

Ionic
Metallic
Covalent

Figure 10-2: The list of elements that form hydrides as they are found on the periodic table. The types of bonding can be ionic, metallic, or covalent.

When hydrogen forms molecular compounds with nonmetallic elements, with the exception of boron, it always forms a single covalent bond. However, catenation can cause the molecules to be quite complex. Catenation involves the formation of chains of elements like in hydrogen peroxide H-O-O-H. This occurs in many organic molecules because carbon has the greatest tendency to catenate.

It should also be noted that all hydrocarbons should be considered as hydrides of carbon.

Metallic hydrides

Not all metals form hydrides, but for those that do, the metallic hydrides are very interesting due to the complex structures they make. They are also important for hydrogen storage because some hydrides can hold and release large amounts of hydrogen, especially when using alloys such as Ni_5La. Currently metal hydrides are widely used in battery technology.

Most metal hydrides can be formed by warming the metal in the presence of hydrogen under high pressure, and it can then be released at a later stage under high temperatures.

Intermediate hydrides

Other metallic hydrides form very complex structures that are not stoichiometric, so are regarded as alloys instead. Take, for example, $TiH_{1.9}$ which is likely composed of $(Ti^{4+})(H^-)_{1.9}(e^-)_{2.1}$. Other examples include $VH_{1.6}$ and $PdH_{0.6}$.

The free electrons are often what account for the shiny luster of the material and give them a very high electrical conductivity. The addition of the hydrogen can cause the material to be more brittle.

In these materials the hydrogen atoms bind in the interstitial sites of the metal lattice and could potentially be used as a material for hydrogen storage.

Applying Itself: Hydrogen's Uses in Chemistry and Industry

Dihydrogen (H_2) is an important industrial chemical, in the petroleum industry it's made from the steam re-forming of hydrocarbon:

$$CH_4 + H_2O \rightarrow CO + 3H_2$$

It can also be made similarly using coke, a fuel with high carbon content, as a reducing agent in a water-gas reaction:

$$C + H_2O \rightarrow CO + H_2$$

Fritz Haber

Fritz Haber is arguably one of the most influential people of the 20th century for his contributions to society. Haber first studied under the famous spectroscopist Robert Bunsen, who you most likely know from the burners that bear his name in the chemistry laboratory or the Muppet bearing his name (Dr. Bunsen Honeydew and his sidekick Beaker). On the positive side of his contributions, shortly after leaving graduate school, Haber co-discovered a process for synthesizing ammonia with Carl Bosch. Many estimates place over 50 percent of the world's current food supply as having been grown with ammonia.

The production of dihydrogen is important, because it doesn't occur naturally on Earth in significant amounts (our atmosphere only has about one hydrogen for every million gas particles present). H_2 is required as a feed stock for many reactions, notably the formation of ammonia:

$$N_2 + 3H_2 \rightarrow 2NH_3$$

The Haber-Bosch process, named after discoverers Fritz Haber and Carl Bosch (see Chapters 16 and 21), is vital for fixing nitrogen to produce ammonia, which is then used for countless applications, though the major use is in fertilizer as a nitrogen-enricher. It works by the direct reaction of nitrogen with hydrogen at high pressures and temperatures, using a catalyst (α-iron).

Deuterium (D) is used commonly in chemistry for analysis using spectroscopy. It's used for the assignment of resonances in infrared (IR), Raman, and NMR spectroscopy. It's also used as a solvent in NMR because D is not observed.

Liquid hydrogen is used in chemistry labs to cool materials to very low temperatures. It's also used as a rocket fuel for high-powered rockets such as those used on the space shuttle. The formation of dihydrogen is exothermic, and this heat can be used for arc welding.

Hydrogen burning in air produces water and heat according to:

$$H_2 + O_2 \rightarrow H_2O + heat$$

In fact, the energy released when hydrogen is combusted is nearly three times greater than that of gasoline or diesel (by weight). Due to this large amount of energy, there is hope that hydrogen could become a viable energy alternative for future generations. Additionally, the lack of CO_2 production (the greenhouse gas produced from burning hydrocarbons) makes hydrogen fuel attractive as a potential environmentally friendly fuel. However, the most common technique of forming hydrogen in bulk is the steam reformation of methane (natural gas), which still utilizes hydrocarbons.

One possibility commonly explored for a "greener" approach to hydrogen production is the electrolysis of water. In short, by running current through water, one can decompose water into H_2 and O_2 (the opposite reaction of burning hydrogen in oxygen). Ultimately, oxygen collects at the anode (where current flows into the water) and hydrogen collects at the cathode (where current flows out of the water). In fact, this electrolysis of water is used on the International Space Station as a method for producing the oxygen the astronauts breathe. The cost of electrolysis, however, is high (which is why it is more cost effective to make hydrogen using hydrocarbons). Also environmentally friendly, hydrogen can be created by certain microorganisms with enzymes called *hydrogenases* that catalyze the oxidation. Hydrogen-producing, genetically modified algae (commonly referred to as algae bioreactors) may prove to be a reliable source of hydrogen in the future, though currently it's still cost-prohibitive.

Whereas hydrogen has a high energy content, it also suffers from having a very low density. This low density results in an inability to transport the same amount of energy as hydrocarbons in the same size fuel tank. Dihydrogen being a gas, requires compression to increase the density; however, that comes at a cost of increasing the strength of the tank, as well as the risks of storing a compressed flammable gas. Some current models of hydrogen vehicles use compressed hydrogen at pressures exceeding 10,000 pounds per square inch (psi).

Overall, hydrogen is a more costly alternative for replacing gasoline and diesel and a more inefficient alternative to electric vehicles, at this time.

Chapter 11

Earning Your Salt: The Alkali and Alkaline Earth Metals

. . . .

In This Chapter

▶ Understanding why Group 1 elements react vigorously with water

▶ Predicting chemical trends of elements with valence electrons in *s* orbitals

▶ Recognizing a diagonal relationship between elements on the periodic table

. . . .

*E*lements in Groups 1 and 2 tend to form ionic compounds. If the ionic compound is the result of an acid-base neutralization, such as when acetic acid (CH_3CO_2H, found in vinegar) neutralizes sodium bicarbonate ($NaHCO_3$, commonly called baking soda), the compound is said to be a salt. For this reaction, the resulting salt is sodium acetate (CH_3CO_2Na); if you neutralize hydrochloric acid (HCl, sometimes called muriatic acid) with sodium hydroxide (NaOH, commonly called lye), the resulting salt is table salt you use on your food (sodium chloride or NaCl). When salts containing Group 1 and Group 2 elements are dissolved into water, they create basic solutions (see Chapter 5 for a review of acid-base chemistry). For this reason, they are sometimes referred to as the alkali (Group 1) and alkaline (Group 2) earth metals, as alkaline is synonymous with basic.

Atoms from both groups have only s orbitals; in Group 1 they have 1 electron in the outer orbital resulting in a +1 oxidation state (or M+ cations), and Group 2 have a +2 oxidation state due to the presence of two electrons. As a result, the reactivity of Group 1 atoms are faster and more furious than those of Group 2 atoms. As a result of the high reactivities, they are never found in metallic form in nature, but are found bound to other atoms such as oxygen.

Elements in both groups are highly reactive and share many characteristics. In this chapter, we describe the Group 1 alkali metals and the Group 2 alkaline earth metals. Specifically, we cover how they react to form various ions and compounds and where you are likely to encounter them.

Salting the Earth: Group 1 Elements

Group 1 includes common elements such as sodium as well as elements that are much more rare (such as francium). The Group 1 elements display the least amount of variation in properties of the compounds they form. Found on the far left side column of the periodic table, all these elements have one electron in the outer orbital that they quite readily give up. This makes these elements much too reactive for them to exist freely in nature. They are highly reactive, combining readily with many other elements.

They most commonly form ionic compounds (see Chapter 8 for details on ionic bonding), or *salts*. This is a result of their tendency to form positively charged ions, or *monovalent M+ cations*. (The exception to this is lithium at the top of the period, or column. With its small size, it has a partial covalent character in many compounds, similar to magnesium in Group 2.) They are called the *alkali metals* because their hydroxides are *alkaline*, or basic, and many of their salts dissolve into water to form alkaline solutions.

All of the metals are very strong reducing agents (Chapter 3 covers reduction reactions). A reducing agent works by giving an electron to another atom. If you rank all metals by their reduction potential, lithium is the lowest and therefore the strongest reducing metal.

The salts of these elements are very soluble in water, due to the low lattice energies and the high hydration energy of the atoms. They all react explosively with water; lithium has the slowest reactivity, and francium the fastest. Cesium is the most electropositive stable element (because francium is unstable), and it's even known to react with ice at temperatures as low as –116 °C (–177 °F).

When reacting with water, these elements form a hydroxide plus hydrogen gas. If you take a piece of sodium metal and dissolve it in water, for example, you form sodium hydroxide (lye) and hydrogen gas according to the following reaction:

$$2\,Na\,(s) + 2\,H_2O \rightarrow 2\,NaOH\,(aq) + H_2\,(g)$$

When elements are heated they give off, or *emit*, waves of light energy known as *electromagnetic radiation*. The color comes from electrons that are being energetically stripped from the orbital in which they reside. As the electron gets or takes energy from its surroundings, it absorbs only the exact amount of energy it needs. Scientists have developed a technique called *spectral analysis* by which to identify elements based on the color of their radiation, or *spectral flame* color. For example, some street lamps are filled with sodium, thus the yellow color due to the characteristic absorption profile of sodium. The other colors they make are shown in Table 11-1.

Table 11-1	Flame Test Colors of Group 1 Metals
Element	*Color*
Lithium (Li)	Red
Sodium (Na)	Yellow
Potassium (K)	Violet
Rubidium (Rb)	Red
Cesium (Cs)	Blue

Lithium the outlier

Lithium is a little different than the other elements in this group. It's considered the *outlier*, or the element that has the least in common with its group members. Apart from sharing the same oxidation number as the Group 1 elements, it's other properties are more similar to Group 2 elements (which are described later in this chapter).

Lithium is a bright silvery metal, first discovered in 1817 by Johan August Arfvedson in Stockholm, Sweden. But it was not until 1821 that William T. Brande isolated the element.

The silvery color of lithium blackens fast when exposed to air because it reacts readily with oxygen (O_2) in the air. This reaction is described by the following equation:

$$4\,Li\,(s) + O_2\,(g) \rightarrow 2\,Li_2O\,(s)$$

It is one of the few elements to react with nitrogen (N_2), so it is used in many reactions involving nitrogen, particularly useful for organic chemists. Breaking the triple bond of N_2 (see Chapter 12 for details on how N_2 is triple bonded) requires energy of 945 kJ/mol. Lithium is able to break this strong bond because a lithium ion has such a high charge density, on account of its small size (see Chapter 2). The following equation describes how the nitride is formed:

$$6\,Li\,(s) + N_2\,(g) \rightarrow 2\,Li_3N\,(s)$$

The nitride can then form ammonia by reacting with water, as described by this equation:

$$Li_3N\,(s) + 2\,H_2O\,(l) \rightarrow 2\,LiOH\,(aq) + NH_3\,(g)$$

Lithium has the most negative standard reduction potential (refer to Chapter 3 if you need a refresher on reduction reactions).

$$Li^+ (aq) + e^- \rightarrow Li(s) \qquad\qquad E° = -3.05 \text{ V}$$

This is one of the most negative standard potentials of any of the elements, which is why lithium is excellent for use in batteries. The lithium ion rechargeable battery used in cell phones and other mobile devices has a lithium cobalt(III) oxide cathode, with graphite as the anode, and an organic solvent as the electrolyte. It works by:

$$LiCoO_2 (s) \rightarrow Li_{(1-x)}CoO_2 (s) + x\, Li^+ (solvent) + x\, e^-$$

At the anode, the Li gets in between the layers of the graphite. This type of reaction is called an intercalation. The Li *intercalates* between the graphite layers and becomes reduced to Li metal. The intercalation results in a change in the structure of the graphite causing it to swell.

Lithium is the least dense of all the metals, it has a density nearly half that of water, this property makes it very interesting for space and aerospace applications, it is used as an alloy, the most common is called LA 141, with a density of 1.35 g/cm^3 it is half the density of aluminum.

Seafaring sodium

Sodium (Na) is the most abundant alkali metal; in fact, it's the sixth most abundant element on earth. Sodium is what makes the oceans salty; therefore, seawater is a good source of sodium (Na^+ ions make up 1.14 percent of sea water).

Sodium chloride (NaCl), is one of the most important chemical compounds ever used. (The word "salary" comes from the fact that payment in some parts of the ancient world was made in salt.)

The element was first isolated using an electrolysis technique by Sir Humphrey Davy in 1806. He extracted elemental sodium by passing an electric current through molten sodium hydroxide. These days, large-scale extraction of sodium is carried out using the *Downs process*, where sodium chloride is molten and then electrolyzed.

Sodium is also an important element for human biology. Each person needs about 1 g of the Na ion on a daily basis. (People often intake up to five times too much daily sodium in their diet.)

Sodium is in high demand for industrial use. It's often used in the extraction of other metals, especially the more rare metals such as thorium, zirconium, thallium, and titanium. It was also used for the production of tetraethyl lead (TEL), which was the famous antiknock (prevented engine knocking) additive used in leaded gasoline fuel from the early 1920s through the 1970s.

To extract titanium, for example, the chloride is reduced with sodium as described by this equation:

$$TiCl_4 \text{ (l)} + 4\,Na \text{ (s)} \rightarrow Ti \text{ (s)} + 4\,NaCl \text{ (s)}$$

The sodium chloride can then be easily washed away using water, leaving the pure titanium metal.

Maintaining your brain with potassium

Potassium (K) is the seventh most abundant element on earth. It's found in silicate rocks and in seawater (potassium makes up about 0.04 percent of sea water). The symbol K comes from the Arabic term kalium, which is an abbreviation of *al qili* (alkaline).

Potassium is absorbed from the soil in plant roots, so it used to be common to boil wood chips to make "potash," which is potassium carbonate (K_2CO_3). In 1807, Sir Humphrey Davy derived pure potassium from caustic potash (KOH), using electrolysis. Potassium is found in the body and is an important element for biological process. About 0.0012 percent of natural potassium is radioactive, and is responsible for some of the heating within your body.

The extraction of potassium on an industrial scale is too dangerous using an electrolytic cell, because of the high reactivity of the metal. For this reason, it is done through chemical means, involving a reaction with sodium and molten potassium chloride described by this equation:

$$Na \text{ (l)} + KCl \text{ (l)} \rightarrow K \text{ (g)} + NaCl \text{ (l)}$$

Potassium is important for the charge balance that occurs in your brain. It helps keep a healthy brain functioning normally and is especially helpful when the brain is under stress. The cells in the brain use the alkali metals to move charge, kind of like electrical current in a copper wire. The charge is balanced in the cells by potassium because it replaces sodium in the cells, preventing them from building up and swamping the cells with too much charge. A good edible source of K can be found in bananas.

Rubidium, cesium, francium, oh my

These elements have similar chemistries to that of the other Group 1 elements, but they are not as commonly found in nature. Going down the period the elements become more reactive and because they are explosive in reactivity, there is not a great deal of chemistry done with them compared with the other Group 1 elements.

Important chemical properties of these elements are:

✔ Francium is the most reactive of the alkali metals.

✔ Cesium-137 is a commonly known radioisotope that forms in the nuclear fission reaction.

Rubidium was first discovered by Robert Bunsen and Gustav Kirchhoff in 1861(this is the same Bunsen of the Bunsen burner that you can find in nearly every chemistry classroom). He was able to identify rubidium because of its unique spectral characteristics. The metal was first produced by the reaction of rubidium chloride (RbCl) with potassium.

Bunsen and Kirchhoff had previously discovered cesium in 1860 in the mineral water of Dürkheim, Germany. It was identified based on the bright blue lines in the spectrum. It was the first element to be discovered by spectral analysis. Cesium metal was first produced in 1882 by the electrolysis of cesium chloride.

Francium was named after France, the country of origin. It was discovered by French physicist Catherine Perey in 1939; she was a student of Marie Curie.

But francium is even more rare as it is only formed from the decay of actinium; all its isotopes are radioactive.

Reacting Less Violently: The Group 2 Alkaline Earth Metals

The elements of Group 2 are similar to the Group 1 elements but are less reactive. They are still difficult to find free in nature because they are reactive enough to form bonds with many elements. They have two outer electrons that are readily donated to form stable compounds. The oxides of Group 2 elements form mildly alkaline solutions so they are referred to as the *alkaline earth metals*.

The Group 2 elements are silvery metals that conduct heat and electricity. They have low first and second ionization energies and form M^{2+} cations (ions with a +2 charge).

Earth, lightly salted

The term *earth* is a tribute to the history of Group 2 metals: The oxides of Group 2 metals were historically called *earths* because they wouldn't dissolve in water or burn in fire. Remember, a *metal* is a good conductor of heat and electricity, so if you were to isolate all the metal ions from a salt into a piece of solid metal it would conduct electrons very well. You're unlikely to encounter Group 1 or 2 metals in metallic form outside of a chemistry laboratory because they are so reactive, so you usually just encounter them as salts.

These M^{2+} cations are much smaller in size then the M^+ cations of Group 1, so Group 2 ions are more strongly hydrated and form larger complexes than the alkali elements. This tendency to form large complexes decreases going down the Group because of increasing ion size, and reduced strength of ion-dipole interactions (see Chapter 8 for details on this).

The alkaline earth metals are not as soluble as the alkali metals because of higher lattice energies associated with the M^{2+} cation. For example, potassium sulfate (K_2SO_4) is soluble in water, but calcium sulfate ($CaSO_4$) is not.

All the M^{2+} cations form complexes with oxygen, nitrogen donor ligands (see Chapter 15), and also with the lighter halides. Beryllium (Be) is limited to a maximum coordination number of four due to its small size (more than four atoms just can't fit around it!). Magnesium and calcium can be six coordinate, and strontium and barium can have even higher coordination numbers.

The alkaline earth metals are malleable and ductile with a silvery luster that's lost upon exposure to air. They have filled s orbitals, are less reactive than the Group 1 atoms, are smaller than the Group 1 atoms, and are more dense as a result. They have two electrons in the outer orbital, making them less reactive than the Group 1 elements, giving them a +2 oxidation state. But all the Group 2 elements act as strong reducing agents because the two electrons of the valence shell can be readily removed, with a decrease in reduction potential going down the Group, but they are less reducing than the alkali metals.

Going down the Group 2, the elements have an increasing Z number due to increasing atomic mass, but with only two outer electrons the atomic radius increases going down the group because of the increasing number of available orbitals; for example 2 s2 orbitals for Be, and 7 s2 orbitals for Ra. They have higher melting and boiling points than the Group 1 atoms because, in the solid state, the electrons are more tightly packed. This explains the higher ionization energies because of increasing energy required to move the valance electrons that are held tight by the nuclear charge. But the two electrons can easily move about in the crystal lattice, resulting in high thermal and electric conductivities.

This also results in them being highly electropositive, the opposite of electronegative, with a decrease in electronegativity going down the period, Be is the most electronegative due to its small size. The small size of Be and magnesium (Mg) gives them a high electronegativity, preventing them from displaying colors upon heating, but the others exhibit colors such as calcium (Ca; light orange), strontium (Sr; brilliant red), and barium (Ba; apple green).

Being beryllium

Beryllium is a rare element that's found mainly in a mineral called *beryl*, which is a complex aluminosilicate ($Be_3Al_2Si_6O_{18}$). When beryl contains chromium impurities, it's known as an emerald; when it has iron impurities, it's an aquamarine.

Compounds of beryllium have a sweet taste but are extremely poisonous. (Unfortunately for early chemists, it used to be common to report on the taste of new materials along with other physical properties.)

Due to its small size, beryllium often forms covalent bonds, making its chemistry distinct from other elements in Group 2. (Magnesium, which we describe next, also forms covalent bonds to some extent.)

The pure element is a steel gray, hard material with a high melting temperature, low density, and high electrical conductivity. It's used in numerous scientific applications such as the windows of X-ray detectors, because of the low Z number (see Chapter 2). Elements with low Z numbers have fewer electrons, and with fewer electrons there is less chance of scattering an X-ray beam. It is also used in the mirrors of space telescopes because of its shiny surface and low thermal expansion.

Most metals form metal oxides when reacting with acids, but when reacting with bases they do not form an oxyanion. These are metal-oxygen compounds with a general formula $M_xO_y^{z-}$. However beryllium is *amphoteric*, which means it reacts with both the hydronium ion and the hydroxide ion, forming an oxyanion, as illustrated in these equations:

$$BeO \text{ (s)} + 2\,H_3O^+ \text{ (aq)} + H_2O \text{ (l)} \rightarrow [Be(OH_2)_4]^{2+} \text{ (aq)}$$
$$BeO \text{ (s)} + 2\,OH^- \text{ (aq)} + H_2O \text{ (l)} \rightarrow [Be(OH)_4]^{2-} \text{ (aq)}$$

Magnificent magnesium

Magnesium was discovered during a drought in England in 1618, when Henry Wicker noticed a pool of water that was being ignored by cattle. He tasted it and found it very bitter. After the water evaporated, he had magnesium sulfate, or Epsom salts.

Magnesium is more abundant in earth's crust than Be and is found in seawater and the mineral magnesite. It's the eighth most abundant element in nature, and it composes 1.14 percent of seawater. Its chloride form is part of the mineral carnalite, and its sulfate form is Epsom salts ($MgSO_4 \cdot 7H_2O$), which you can find at your corner drugstore. It's also found in asbestos ($CaMg_3(SiO_3)_4$) and talcum powder ($H_2Mg_3(SiO_3)_4$).

Magnesium forms compounds with covalent bonds, making its chemistry a little different from the Group 2 elements below it in the Group.

When Mg is ignited, it burns with a bright white light that can harm the retina of the eye. It was used to create a bright flash for early photography. The flash was created by combustion of magnesium powder described by this equation:

$$2\,Mg\,(s) + O_2\,(g) \rightarrow 2\,MgO\,(s)$$

The resulting fire from this reaction is so vigorous that even a fire extinguisher using CO_2 can't be used because it'll further combust the CO_2 as described here:

$$2\,Mg\,(s) + CO_2\,(g) \rightarrow 2\,MgO\,(s) + C\,(s)$$

Magnesium can also react with organic compounds to form organomagnesium compounds, called *Grignard reagents*.

Grignard reagents are used extensively in organic chemistry to assist in the formation of C-C bonds. For example, magnesium metal reacts with halocarbons (or alkyl halides) such as bromomethane, in a solvent called ether. The magnesium inserts itself in between the carbon and halogen atoms, and each forms a covalent bond with magnesium.

$$C_2H_5Br\,(ether) + Mg\,(s) \rightarrow C_2H_5MgBr\,(ether)$$

Magnesium is used industrially as an alloy material with aluminum. They both have low density and decrease the total weight of whatever object they are used to make. They were used in naval vessels to decrease weight and increase speed; however, when hit by a missile or torpedo, the alloy was flammable — again demonstrating the high reactivity of the alkali earth metals.

Commonly calcium

Calcium metal was first obtained in a pure form through the use of electrolysis in 1808 by Sir Humphrey Davy. But calcium, in the form of quicklime (CaO), had been used for thousands of years before this to neutralize acidic soils.

Calcium is the fifth most abundant element in the earth's crust, comprising about 3 percent of it. It's most often found in limestone deposits. Its carbonate ($CaCO_3$) is found in many different forms with varying physical properties, from soft chalk to very hard marble. It's also found in other common materials such as gypsum($CaSO_4 \cdot 2H_2O$) and in your bones as hydroxyapatite ($Ca_5(PO_4)_3(OH)$). Calcium can form oxides, hydroxides, and halides. The carbonate is used as an anti-acid to neutralize stomach acid. The hydroxide form is used as cement, and the chloride form is used for delicate sculptures in the form of alabaster. In the sulfate form it makes gypsum casts that are used for plastering walls because of its fire-retardant properties. The carbide form is used as a source of light; this forms the basis of acetylene lamps.

Calcium ions (Ca^{2+}) play an important role biologically, particularly in teeth and other structural bio-minerals (such as the apatite in your bones). Unlike Be, which is used as a window in X-ray machines, the calcium ion is very dense material and is picked up within an X-ray image (because it blocks the X-rays, leaving an outline of your bones on the X-ray film).

Calcium is a grayish material that reacts slowly with oxygen to give the oxide:

$$2\,Ca\,(s) + O_2\,(g) \rightarrow 2\,CaO\,(s)$$

Calcium oxide, or quicklime, is made in large quantities within the steel industry. It is formed by heating calcium carbonate at 1170 °C:

$$CaCO_3\,(s) \xrightarrow{\Delta} CaO\,(s) + CO_2\,(g)$$

The oxide has a high melting point, and when heated with a direct flame it glows a bright white color. Before electric lighting, theaters exploited this thermoluminescence to light their stages, hence the term "to be in the limelight."

Strontium, barium, radium

Strontium, barium, and radium are much more rare than other Group 2 elements. They exhibit all the properties of Group 2 elements listed earlier but are less commonly found in nature or used by industry.

Strontium was discovered by Irish chemist Adair Crawford in 1790 when he was studying the mineral witherite ($BaCO_3$). But it's named after the Scottish village Strontian, where it was discovered in lead mines in 1787; it was not until 1808 that Sir Humphrey Davy isolated it for the first time. It was isolated using a mixture of strontium chloride and mercuric oxide in an electrolysis reaction. ^{90}Sr is formed in the nuclear fission of ^{235}U and ^{239}Pu; unfortunately it is a very toxic radionucleotide, and levels have increased globally due to nuclear weapons testing. After nuclear testing, the physical material deposits on the surface of the earth and is ingested by land mammals where it can replace calcium. It was found in large supplies in cow's milk, which ultimately led to a ban on nuclear testing. When heated in a flame, it gives off a brilliant red color that is often utilized in fireworks.

Barium is found in the mineral barite, first discovered in the Italian university town of Bologna. The mineral would continue to glow after irradiation by light. Sir Humphrey Davy isolated Ba^{2+} in 1808. It's used mostly in the oil industry in the form of drilling mud. Due to the heavy weight of the atoms, it can help prevent oil from exploding out of the well. It gets its name from the Greek word *barys*, meaning "heavy." It has a large Z number and can be used as an effective X-ray tracer. It can be ingested by patients to work out where a kidney stone resides in the body, for example, because the barium will not pass the stone and the blockage can be seen using X-rays.

Radium was discovered by Marie Curie and her husband Pierre in 1898 while removing uranium from pitchblende. They noticed that the remaining material was still radioactive. It wasn't until 1910 when it was isolated as a pure metal; Marie Curie and André-Louis Debierne achieved this by electrolysis.

Diagramming the Diagonal Relationship

The diagonal rule is a useful tool to predict the properties of atoms that lie diagonally from one another on the periodic table. As you go down the Group, the atomic radius gets larger; however, as you go to the right across a period, the atomic radius gets smaller. The trend of properties canceling out is seen for electronegativity as well: Going down the Group causes electronegativity to decrease, but going right causes electronegativity to increase. For these reasons, there's a strong diagonal relationship between lithium and magnesium.

So if you go down diagonally from one column to the next , the atomic size remains approximately equal. The diagonal relationship is due to a combination of atomic size and charge.

The result is that the first element of the main groups (lithium, beryllium as well as boron, carbon, nitrogen, oxygen, and fluorine described in Chapter 12) are different from the rest of their group members. Usually, the first element is much smaller than the rest of the elements in that group, and it tends to form covalent compounds and complexes. This tendency is called *Fajans' rule* after Kazimierz Fajans, a physical chemist who in 1923 postulated it.

As well as the obvious horizontal and vertical trends found in a periodic table, there is also a diagonal relationship between certain sets of elements (lithium and magnesium, beryllium and aluminum, boron and silicon). The diagonal relationship relates an atom to another one that is one over, and one down from it. The effect of Fajans' rule means that pairs of atoms are found to have similar size and electronegativity, and this leads to similarities in reactivities and other properties.

It works because changes in the atomic radius and the ionic radius cancel each other out going down diagonally. There's an increase in ionic charge going down the table, and a decrease in ionic size going left to right in the table. The coupling of ionic charge and ionic size affect the polarizing power of the atoms. It's the change in polarizing power that's cancelled out when going diagonally down the table; this is Fajans' rule.

Chapter 12

The Main Groups

*T*his chapter introduces you to the six main groups of elements: the boron group, the carbon group, the nitrogen group, the oxygen group, the halogens, and the noble gases. There are a total of 37 elements in these six groups, but this chapter focuses on only 19 of them that best represent their group characteristics.

The most interesting point about the elements described in this chapter is the variety of properties that they exhibit. These six groups include elements that are nonmetals, metalloids, and metals, as well as elements that are solids, liquids, and gases at room temperature.

You may be surprised to find out that you encounter many of these elements on a daily basis. Some of them are important for industry (such as silicon used in electronics) whereas others, such as carbon and nitrogen, are important for biology. Also, don't forget the chlorine in swimming pools and helium in balloons.

The chemistry of these six groups of elements is made similar by the presence of occupied *p*-orbitals, which control much of the chemistry and material properties. For example, the difference between diamond and graphite can be understood in terms of the presence of sp^3 (this is read aloud as s-p-three) carbon atoms found in a diamond compared to sp^2 carbon atoms found in the graphite of pencil lead (it's not actually made of lead). The sp^2 carbon atoms have one unbonded electron in the outer shell, giving graphite the capacity for conduction of electricity because that electron can flow more freely, compared the fact that all the electron in diamond crystal are used for covalent bonding.

Placing Main Group Elements on the Periodic Table

The periodic table is built on a simple idea: The activity of atoms is related to the number of valence electrons (electrons capable of participating in bonding; for more on valence electrons, check out Chapter 7) in the outer orbitals. All of the atoms in the same column (or group) on the periodic table have the same number of valence electrons in their outer orbital, and, therefore, exhibit similar behavior when reacting with other elements.

The portion of the periodic table that we cover in this chapter is shown in Figure 12-1. Notice that the alkali and alkali Earth metals aren't included in this list, although they are also considered main group elements, they were treated on their own in Chapter 11.

Here are two properties that can help you understand these elements in almost every case:

✔ Electronegativity is the tendency of an atom to attract electrons to itself (as a result of the combination of *valency* and *orbital symmetry;* see Chapter 7 for more details on these topics). It mainly affects the reactivity of the main group metals. As you move to the right on the periodic table, the atoms are more electronegative, so they have a stronger desire to attract electrons.

✔ Ionization energy reflects that both the first and second ionization energies (the energy needed to move an electron), affect the type of structures and compounds that can be formed by a particular element. (See Chapter 2 for more details on ionization energies.)

		Pnictides	Chalcogen	Halogens	Noble Gasses
					He
B	C	N	O	F	Ne
Al	Si	P	S	Ci	Ar
Ga	Ge	As	Se	Br	Kr
In	Sn	Sb	Te	I	Xe
Ti	Pb	Bi	Po	At	Rn

Figure 12-1: Main group elements on the periodic table.

Non Metals	Metals	Noble gasses

When looking at each element, it may help to think about how that group forms bonds with other elements. Here are a few simple questions to ask yourself that can help you understand how each element behaves:

- ✔ How well does the element bond to itself? For example, C-C bonding versus F-F bonding.

- ✔ How well does it bond to hydrogen? Does it make hydrides, and, if so, what kind of interesting properties can we observe?

- ✔ How does it bond to the halogen group? (The halogens bond to nearly every single atom, and so they are often used as a comparison to demonstrate material properties.)

- ✔ How does it bond to metals? And if it is a metal atom, how does it bond to nonmetals?

- ✔ How does it react to oxygen and carbon? Both oxides and carbides form a large class of materials in and of themselves.

Ask yourself these simple questions as you look at each group (and then each element), and you start to see the larger picture of how the elements are related to one another.

Lucky 13: The Boron Group

Group 13 elements, also called *the boron group,* include a wide variety of properties. Some of the important properties about this group include:

- ✔ They readily form Lewis acids, three coordinate compounds that are capable of accepting an electron pair and increasing the coordination number.

- ✔ All the Group 13 elements have the outer configuration of ns^2np^1. (Chapter 2 covers electron configurations if need a refresher.)

Most of the boron group elements exhibit a tripositive (3+) oxidation state; however, they can be occasionally found in a unipositive (1+) state (with the exception of boron itself, which we describe in more detail later in this chapter). Keep reading to find out the details of five of the Group 13 (the 13th column on the periodic table) elements: boron, aluminum, gallium, indium, and thallium.

Not-so-boring boron

Boron (B) can be identified by its characteristic green flame, which was first reported by Geoffroy the Younger in 1732 but it was not until 1808 that elemental boron was first crudely isolated by three different scientists:

Louis-Joseph Gay-Lussac and Louis-Jacques Thenard in Paris, and Sir Humphry Davy in London. Boron is difficult to isolate, and when the purest form was isolated by E. Weintraub in 1909, it was found to have wildly different properties of the impure boron samples synthesized prior to 1909.

Boron has a unique chemistry that's dominated by *electron deficiency,* which means it doesn't form covalent bonds often. The structures that boron makes are complex, particularly when bonded with metals. The following are some of the most important characteristics of boron.

- Boron is a hard, brown-black powder.
- Boron is the only nonmetallic element in Group 13.
- Hydrides of boron consist of a vast range of neutral compounds and anions; with the exception of BH_4^- ion, these compounds exhibit very complex structures.
- Halides of boron are Lewis acids, with an acceptor strength that follows the order: $BI_3 > BBr_3 > BCl_3 > BF_3$.
- Oxygen compounds of boron such as B_2O_3, $B(OH)_3$, and $Na_2B_4O_7$ (borax) can form complex structures with existing metal borates that contain both 3-coordinate and 4-coordinate boron atoms.
- Some boron-nitrogen compounds form similar structures to carbon compounds. Adding boron or nitrogen to carbon compounds (such as diamond) can alter the physical properties of the diamondoid, making it more conductive for example.

Boron is renowned for making clusters, particularly with hydrogen. In the chemistry lab it's often found as a hydride, or a species with an anion of hydrogen (H^-) such as sodium borohydride, $NaBH_4$. Amorphous (solid) boron is used as an igniter for rocket fuels. It's added to liquid hydrogen to increase the rocket fuel efficiency. Because boron can hold and release so much hydrogen, it has also been touted as a valuable material for the developing hydrogen economy.

The largest cache of boron is found in Death Valley, California, in a mineral called borax (sodium tetraborate, $Na_2B_4O_7$), but it can be found in other parts of the world such as in volcanic spring waters and places such as Tibet, China, and Saudi Arabia.

Borax is a common household cleaning product, and is also found in the household as boric acid (H_3BO_3) that can be used as an antiseptic. It is also found in Pyrex dishes, which are made of boron-aluminum-silicate.

Boron and gallium (discussed more in depth later in this chapter) are both used as a *dopant* (a material added to alter a material's electronic properties) in semiconductors.

When as semiconductor material is not doped, but is made of pure silicon, it is known as an *intrinsic* semiconductor. But when there is some addition of another material, such as boron or gallium, it is referred to as an *extrinsic* semiconductor.

Gallium is also used along with arsenic to make nanoparticles of gallium arsenide that are commonly referred to as *quantum dots*. They have unique optoelectronic properties and are useful for solar cell technology as well as imaging in biological systems.

An abundance of aluminum

Even though it's the most abundant metal on earth, aluminum (Al) is usually tied up in oxide compounds. It wasn't until a Danish scientist, Hans Christian Ørsted isolated it in 1825 from aluminum chloride($AlCl_3$) that pure metallic aluminum was found. At the time, it was so rare that it was a valuable metal. Napoleon was said to show off to state dignitaries by allowing them the honor of using his aluminum cutlery, which at the time was more precious than either silver or gold. When the technique for separating the metal was perfected in the late 19th century, the price of Al fell nearly a thousand fold, and the use of aluminum sky rocketed.

Elemental Al is silvery in color and luster, but is found in minerals such as bauxite and in various clays (not commonly known for their shininess).

Important properties of aluminum include the following:

- Aluminum oxide is amphoteric and can react as either an acid or a base.
- Aluminum valence is +3.
- Aluminum is a metal that is both ductile and malleable.
- Aluminum has a high reduction potential, causing the extraction process to be energy intensive.
- Aluminum has high strength with a low density of 2.7 g/cm^3.

The modern world has changed significantly as a result of the many applications for aluminum. From the wrap on your sandwich to the hulls of space ships and airplanes, it's the material of choice for many common items.

Mendeleev's Missing Link: Gallium

Gallium (Ga) was predicted to exist before it was ever isolated. Dmitri Ivanovich Mendeleev predicted an element as a missing link between aluminum and indium, and at the time he termed it ekaaluminum. It was later discovered in 1875 by French chemist Lecoq de Boisbaurden using a technique called spectroscopy. (A scientist using spectroscopy looks at the light released from an atom that is excited. The color of a flame, for example, or light that is emitted from a hot material indicates which elements are present.) He named it gallium, based on the Latin word *Gallia* for France.

Gallium is widely found in nature in minerals such as gallite, bauxite, and germanite. It's found in all aluminum ores. Gallium is a gray orthogonal crystal, or a silvery liquid. Other important chemical properties of gallium are as follows.

- ✔ Gallium is found with all minerals of aluminum and is a by-product of aluminum production.

- ✔ Gallium forms mostly binary and oxo compounds in a +3 oxidation state; for example, gallium forms a stable Ga_2O_3, and a volatile suboxide of Ga_2O.

- ✔ Gallium melts near room temperature (just 29.6 °C) and remains a liquid through a large range of temperatures (up to 2,204 °C).

Increasing indium use

Indium (In) was first isolated by Ferdinand Reich in 1836. However, he was color blind and confused the brilliant indigo spectral color of indium with thallium. Indium was later identified by Hieronymus Richter in 1863 as part of ZnS, a mineral called zinc blende.

Important properties of indium include the following.

- ✔ It's a rare element with low abundance on earth.

- ✔ It's recovered from zinc ores after the extraction of zinc whereby the residues are treated with a mineral acid.

- ✔ Indium is stable in air at ambient temperature, but oxidizes under high temperatures to form indium trioxide, In_2O_3.

- ✔ The most common valence state is +3, but compounds of +1 are also commonly obtained.

- ✔ When a piece of indium metal is bent, it gives off a high-pitched shriek.

- ✔ Indium-115 has a large neutron cross section area, and can capture large amount of neutrons; thus it's used in nuclear plants to determine the activity of nuclear reactors.

Despite being a rare element, indium is used extensively today, in particular in the form of indium tin oxide (ITO) used in touchscreen displays. The price of indium has skyrocketed in the last decade because of the new found application as a transparent conductor; its value is expected to increase as its limited supplies dwindle.

Toxic thallium

British chemist Sir William Crookes first discovered thallium (Tl) in 1861 while doing spectroscopy on tellurium. He saw beautiful green lines in some leftover waste material from a factory that made sulfuric acid and named the element after the Latin word *thallos,* meaning budding green twig. It was later isolated in 1862 independently by both Sir William Crookes and Claude-Auguste Lamy.

Some things to know about thallium include the following:

✔ Thallium has two naturally occurring isotopes — ^{203}Tl and ^{205}Tl — yet it has 28 artificial radioisotopes, with half-lives ranging from 3.78 years to 0.2 seconds.

✔ It's a toxic material that can cause serious poisoning if ingested and has been used in insecticides and rodenticides.

✔ It forms all its compounds in two valence states: +1 (thallous), and +3 (thallic).

Thallium occurs in nature in potash minerals and pyrites from which the metal is recovered. Thallium has a metallic luster when freshly cut, but upon exposure to air it forms a bluish-gray appearance that is similar to lead. It combines with several elements forming binary compounds.

It's used in thallium-mercury alloys for applications in switches that are used at sub-zero temperatures, and it can be used in laboratories as a low temperature thermometer because it extends the freezing point of Hg to –60 C. This can be used by engineers to monitor conditions at such low temperatures.

The Diamond Club: The Carbon Group

The carbon group include important and well-known elements such as carbon, silicon, lead, and tin. The properties and chemistry of this group resemble the boron group because both groups include metal and nonmetal elements. Carbon and silicon are nonmetals whereas germanium, tin, and lead are metals. The metallic properties increase going down the group (or column).

All members of the group have an oxidation state of +4 but with the capacity to form elements in a +2 oxidation state. The stability of such +2 compounds increases going down the table; for example, carbon dipositive molecules (also called *carbenes*) exist only as intermediates between more stable states. But for lead and tin, the lower oxidation state of +2 is the most stable.

All of the elements bond readily with hydrogen to form *hydrides* (named according to what atom is at the center such as alkanes with carbon at the center, silanes with silicon, and germanes with germanium). Bond strengths for each element decrease down the column with silicon forming the strongest bonds and lead forming the weakest. These elements can also bond to themselves to provide a backbone flanked by hydrogen atoms to create a hydride molecule. The stability of these hydride molecules follows C > Si > Ge. The alkanes are double-bonded and therefore are more stable and more numerous. The halides of carbon, silicon, and germanium have similar structures and formulae. Silicon and germanium halides are Lewis acids that are readily hydrolyzed by water.

In a sense, humans are members of the diamond club because carbon (which is what diamonds are made of) is found in nearly every material we encounter, including the materials of which we are made. Living matter is made of carbon molecules and carbon-containing compounds. And since World War II, carbon has been put to so many uses that it would be impossible to imagine modern life without it.

Silicon is another wonder element of the 21st century due to its use in the semiconductor industry; germanium, tin, and lead are not as well known or used as carbon and silicon, but have also found useful applications in industry. This section explains the specific details of each of the carbon group elements.

Captivating carbon

Burn a campfire and in the morning, all that's left is carbon (C). The ancient Romans collected the soot of burned pine branches and used it to make inks that lasted centuries — long enough to tell us the story of ancient Roman life.

There are more known carbon compounds than any other element, except maybe hydrogen. Carbon even has the honor of an entire branch of chemistry developed around it. *Organic chemistry* is the science of carbon. Originally named based on the fact that all living and most organic matter is composed primarily of carbon, organic chemists do not always deal with the chemistry of life; however, the name has stuck nonetheless. In fact, the distinction between organic and inorganic is simply that carbon bonded to a metal is typically considered an inorganic compound whereas carbon bonded to nonmetals is considered an organic compound. Organic chemistry has been applied to make polymers, plastics, drugs, and a whole slew of other molecules we consume or utilize on a daily basis.

Carbon is found as three *allotropes,* or structural forms: diamond, graphite, and fullerenes. Even though they are made from the same atom, these three allotropes have wildly different properties and applications because of the different bonding arrangements within each one.

Some important chemical properties of carbon you should keep in mind include the following:

- Carbon has the ability to undergo *catenation,* which means it can form long, chained molecules of carbon atoms strung together.

- Carbon has a valence of 4 and as such can bind with up to four other atoms.

- Carbon can have a coordination number of 2 ($\equiv$C– or =C=), 3 (=C<) or 4 that have either linear, triangular (planar), or tetrahedral geometries, respectively.

- Carbon-carbon bonds are very stable, and so carbon compounds can have very high strength.

The number of compounds made with carbon is staggering, and the various physical properties they possess range from soft to hard, opaque to shiny and even clear. Carbon compounds conduct electricity and heat, as well as insulate heat and electricity. There are so many uses for this element that some have suggested humans have entered into the "diamond age."

Coming in second: Silicon

Silicon (Si) is the second most abundant element on this planet (oxygen takes first place in planetary abundance). The name *silicon* comes from the Latin words *silices* or *silex,* which means flint. Flint is a type of rock made of silica (SiO_2) that can be shaped into stone tools such as spear heads or axes, and was used by ancient humans all over the world for this purpose. It was not until 1824 that elemental silicon was discovered. At the time, it was named *silicium* but was later changed to silicon. This illustrated the fact that it was not a metal but sounded similar to the other nonmetal elements such as carbon, boron, nitrogen, oxygen, and so on.

Important chemical properties of silicon include the following:

- Silicon forms mostly covalent compounds.

- Silicon also forms very stable oxides.

- Similar to carbon, silicon has a valency of 4.

- Unlike carbon, with coordination numbers ranging from 2-4, silicon's coordination numbers range from 2 to 6, meaning silicon is capable of being *hypervalent,* or having an expanded octet of electrons.

- Silicon has a lower electronegativity than hydrogen, whereas carbon has a higher electronegativity than hydrogen.

Silicon is usually found as an oxide in various forms of silica (SiO_2) and silicates (SiO_4^{4-}), but it is never found free in elemental form in nature because it binds readily with widely available oxygen and hydrogen. It's found in quartz minerals and sand grains, and used in several industries including cement manufacture and information technology. After carbon and hydrogen, silicon forms the highest number of compounds.

Germane germanium

Germanium (Ge) was predicted to exist by Mendeleev, and was discovered just 15 years later by Clemens A. Winkler in Freiberg, Germany in 1886 (hence the name germanium, in tribute to the country of discovery). It was first encountered as an ore from a deep silver mine in Freiberg. Germanium is a metalloid (exhibiting some metal properties) and is more reactive than silicon; it can be dissolved in both sulfuric acid and nitric acid. Germanium dihalides are stable.

Germanium is less common than carbon or silicon and less well known than tin or lead. Yet it is used extensively in all fiber optics, and it is used in some semiconductors too. It used to play a major part in the electronics industry as a rectifier because it could transmit electricity in one direction but not in the reverse direction. It was the element that was first used to make a transistor and was used in the first commercial transistor radios. But today it is less common to find it because it has been replaced by the cheap and prevalent silicon. It's still used in specialty devices such as wide-angle camera lenses, detectors, and some alloys.

Malleable tin cans

Tin (Sn) is commonly found as a mineral called cassiterite (SnO_2). For thousands of years it has been used as an alloy in bronze to build weapons and tools. Today it is most commonly found in the tin cans that hold the peas and peaches you buy at the grocery store. It's a silvery-white metal that is malleable as well as somewhat ductile. It melts at 231.9 °C, so it's easy to purify. It reacts with carbon at high temperatures to obtain the pure metallic form.

The most important compounds of tin are tin(II) fluoride (SnF_2), which is used in toothpaste, and tin(II) chloride ($SnCl_2$). Tin also reacts with all halogens to form their halides. *Alkyltin,* which is an organometallic compound, is made on a large scale and is one of the only examples where it is not used commercially as a metal.

Even though it's stable in air at room temperature, the powder form oxidizes readily. When burned, the tin oxide gives off a green flame.

Plumbing lead

Lead has the symbol Pb, from the Latin word *plumbum*. For thousands of years it's been used to make water pipes (thus the term *plumbing*). However, it's considered poisonous, and although it can still be found in paint and other materials, it's no longer commonly used for household pipes. Lead is one of the oldest materials known to civilization and has been mined for over 6,000 years.

Lead is found in nature in several minerals, but never in pure metallic form. Lead forms amphoteric compounds in +2 and +4 valence states. There are over 1,000 compounds of lead known. Most lead salts are only slightly soluble in water, and halides of lead are always anhydrous.

The main use for lead today is in solder, bullets (it is one of the heaviest stable elements), and car (lead-acid) batteries. Because it's so malleable and soft, yet not prone to rusting (oxidation), it was thought to be excellent for piping until its toxicity was understood. Lead can still be found in some paints because it creates certain colors such as chrome yellow (lead chromate), white (lead sulfate), and red (lead tetroxide). Lead was also widely used as an additive to automotive fuel in the form of tetra-ethyllead, but it is not as common anymore.

Today, lead is usually purified from an ore called galena (PbS). Lead is so heavy it can be separated from other metals by gravity separation and flotation methods (see Chapter 17 for details on these methods); however, in the oxide form, it's reacted with carbon to get pure metallic lead.

Noting Pnictides of the Nitrogen Group

The Group 15 elements, or *pnictogens* or *pnictides* (a term which refers to their capacity to suffocate), are commonly referred to as the *nitrogen group*. The key trends of this group are not as straightforward as the other groups. This group poses some interesting problems when trying to find similarities and differences as the atomic numbers increase. A few things to keep in mind about this group are as follows:

✔ As mentioned earlier in this chapter, there is an overall increase in the metallic character of the elements going down the period.

✔ Ionization energies increase sharply upon the removal of the p electrons.

✔ Enthalpy of atomization decreases steadily going down the group (similar to the boron and carbon groups described earlier).

✔ Nitrogen is a typical nonmetal, but bismuth at the end of the group is a typical metal.

✔ The ionization energies decrease steadily as the atomic number increases going down the period (or column) in the nitrogen group.

✔ The nitrogen group forms many oxides, sulfides, halides, and hydrides.

Leading the pnictides: Nitrogen

Nitrogen (N) is a nonmetallic element that occurs as a diatomic gas (N_2). Nitrogen gas is 78 percent of the air you breathe every day. Considering its abundance, it was discovered as an element relatively recently in 1772.

Nitrogen is found commonly as *nitrates*. It's a vital element to living systems, and nitrates are often used as plant fertilizers. Nitrogen is an important nutrient that cycles through ecosystems. Until it was possible to make bioavailable nitrogen in large quantities, wars were fought over nitrogen sources.

Important chemical properties of nitrogen include the following:

✔ Nitrogen is a colorless and flavorless gas that converts to liquid at −195.79 °C. Liquid nitrogen is used as a coolant in many applications because of this.

✔ Nitrogen tends to form multiple bonds (such as dinitrogen in the air that has a triple bond that makes it inert).

✔ All of the oxides of nitrogen can be thought of as strong oxidizing agents.

✔ Halides of nitrogen can form, with fluorides being the most stable.

✔ Several nitrides (3− anions) are also known, but they are not highly ionic.

Commercially, nitrogen is used to produce ammonia (NH_3) using the *Haber-Bosch process*:

$$N_2 + 3H_2 \rightarrow 2NH_3$$

The development of the Haber-Bosch process, in many ways was the key chemical process of the industrial age. With it the nitrogen supply soared, and this was mainly put to use in agriculture, thus leading to the large-scale food production that fed a booming population.

Ammonia is an important industrial and laboratory chemical; it is basic in water, and acts as a good ligand. Similar compounds such as amines are also an important part of biology. For more information on ammonia, see the sidebar on Fritz Haber in Chapter 10.

Nitrous oxide (laughing gas) is still used in some dentist's offices to relieve anxiety during dental procedures. And hydrazine, N_2H_4, is used as a rocket fuel in the space industry.

Finding phosphorus everywhere

Phosphorus (P) is one of the most widely distributed elements on Earth. It's found as phosphate in almost all igneous and sedimentary rocks. It's also an essential element for life and provides energy when adenosine triphosphate (ATP) is converted to adenosine diphosphate (ADP) within the cells of your body. It can be found in bones, in DNA, and nucleic acids. Today, it can be found frequently in soda drinks in the form of phosphoric acid. On its own, however, phosphorous is a hazardous material, and prolonged exposure can lead to severe burns and illness.

Important chemical characteristics of phosphorus include the following:

- ✔ Some halides of phosphorus are good Lewis acids. Most of these compounds have a +3 or +5 oxidation state.

- ✔ The oxides of phosphorus are also typically found in the +3 and +5 oxidation states; these compounds are often polymeric.

- ✔ Phosphorus compounds with metals are known and have a low ionic character.

Melding the metalloids: Arsenic and antimony

Whereas nitrogen and phosphorus were both nonmetals, arsenic (As) and antimony (Sb) are both metalloids. Although phosphorus is normally found in minerals, arsenic and antimony are more often found as sulfides in nature and in significantly smaller quantities. Pure samples of each metalloid are usually dark gray in color.

One of the most common applications for these two metalloids is for use when strengthening alloys, or mixtures of elements, especially lead. Lead is a relatively soft, dense metal, so to make lead harder for applications such as bullets, either metalloid can be added to help make the resulting product stronger. This practice of alloying arsenic and antimony with lead dates back to the Bronze Age.

Another use for arsenic is as a poison, sometimes for good in the cases of pesticides and insecticides and sometimes for bad in the cases of chemical warfare and homicides. As a chemical warfare agent, arsenic's role as a rainbow herbicide is most likely the most horrific. Similar to the well-known Agent Orange, Agent Blue (another rainbow herbicide) was a chemical used in the Vietnam war to destroy plants, serving to both destroy the enemies' food supply and any cover the plants may have provided for the soldiers. Agent Blue included cacodylic acid $((CH_3)_2AsO_2H)$, an acid that has arsenic as a central atom.

Arsenic poisoning is a serious concern in areas with naturally contaminated water supplies, such as Bangladesh's ground water. Some estimates are as high as 20 percent of all wells around the world suffer some sort of arsenic contamination, which can lead to both skin and bladder cancers (and ultimately death).

A mixture of chromium, copper, and arsenic (known as CCA) has been used for many decades as a wood preservative; you can usually tell if wood has been treated by CCA as it will have a light-greenish tint. The use of CCA as a preservative is slowly being phased-out, however, due to concerns that acidic environments (such as acid rain, for example) could cause the arsenic to leach out of the wood and into the environment.

The hydrides of arsenic and antimony are less stable than ammonia and are also less basic. But there are still a number of organic compounds that can be made with these elements, such as trimethylarsine.

Keeping Up with the Chalcogens

The chalcogen group of elements includes oxygen and sulfur, important elements for life, as well as selenium, tellurium, and polonium. The name stems the Greek word *chalkos,* meaning "bronze" or "ore," and *genes,* which means "born."

Even though these elements are in the same group as oxygen, they are remarkably different — they are much less electronegative and the hydrides of chalcogen elements other than oxygen, are poisonous gases.

All of the chalcogens form *chalcogenide* 2⁻ anions such as *oxides, sulfides, selenides,* and *tellurides.* Selenium is used in photocopiers, it makes the electrostatic charge on the paper that directs where the ink sticks. Polonium is the most unique element in this period since it is a metal. Keep reading to find out the important properties of oxygen and sulfur, two very important chalcogen group elements.

Oxygen all around

Oxygen (O) is the most widely abundant element on Earth, making up 50 percent of the Earth's mass. You probably know about its presence in the air you breathe, but it also occurs as a solid in various oxides, silicates, hydroxides, nitrates, sulfates, and carbonates. Even the moon, with no air to breathe, has an abundance of oxygen in the form of solid oxides.

The name *oxygen* reflects the Greek *oxys* meaning "sharp taste," and *gonos* meaning "former." At the time, it was named by French nobleman and chemist Antoine Lavoisier, he thought it was an essential component of all acids. This was later proven untrue, but the name stuck.

Elemental oxygen can be created by the thermal decomposition of many metal oxides. The earliest reactions involved the reaction of mercury(II) oxide:

$$2\,HgO\,(s) \;\rightarrow\; 2\,Hg\,(l) + O_2\,(g)$$

Joseph Priestly was one of the first to isolate oxygen. He achieved this using mercury oxide in 1774 by harnessing sunlight as an energy source.

Some things you should know about oxygen include the following:

✔ Oxygen is a colorless and odorless gas.

✔ Oxygen reacts with almost all elements.

✔ It has a valency of 2 (it's divalent).

✔ Catenation is possible; for example, in hydrogen peroxide is H-O-O-H.

✔ In liquid form oxygen is a stunning blue color, but it's also highly explosive. (Liquid oxygen was used as a fuel for the former space shuttle because of its highly oxidizing nature, which can make it react quite explosively.) The gaseous form of oxygen at sea level is a diatomic molecule, O_2. At higher elevations gaseous oxygen also occurs as ozone, O_3 that is formed by radiation from the sun. At even higher altitudes it is found as atomic oxygen.

Although some oxygen reactions are exothermic (giving off heat), others require the addition of heat to reach a final product. Some of the earliest developments in modern chemistry arose due to the investigation of combustion reactions that involved compounds of oxygen. When heated with oxygen, all organic molecules undergo combustion reactions such as:

$$2\,C_4H_{10} + 13\,O_2 \;\rightarrow\; 8\,CO_2 + 10\,H_2O$$

Oxygen can also be corrosive due to its high reactivity. This can be detrimental because metals that oxidize readily lose some of their strength as a result. An example is when iron is oxidized and turns to rust.

Sulfur

Elemental sulfur (S) has been known since ancient times, its name coming from both Sanskrit (sulvere) and Latin (sulfurium). In 1808, a sulfur dioxide sample was burned to release the oxygen, but it was found to contain hydrogen as a contaminant. Eventually it was truly purified and was then verified to be an element as suggested by French chemist Antoine Lavoisier.

Sulfur is found in free elemental form in nature (as volcanic brimstone) and in minerals such as gypsum ($CaSO_4 \cdot 2H_2O$) and pyrite (FeS_2). These minerals are often found in deposits underground, and the sulfur can be extracted using the Frasch process, which was developed in the Gulf coast region of the United States to access large deposits of sulfur available in that region.

Chemists often use the prefix *thio* to denote the presence of sulfur in a compound, such as thiophene (C_4H_4S).

Important chemical properties of sulfur are listed here.

- Sulfur exhibits *catenation,* the ability to form long chain molecules.
- Sulfur forms allotropes that contain both chain and ring structures.
- There are several oxides known with formulae such as EO_2, and EO_3. There is a tendency to make more polymeric structures with the heavier elements.
- Sulfur halides have a range of oxidation states up to +6 — most of them are molecular compounds, but some have polymeric structures.
- Sulfur acids easily oxidize metals and can be used for cleaning heavy metals.
- Sulfur forms strong bonds with metals, such as gold.
- Sulfur can be used as an inhibitor in catalysis (to slow down the reaction).

Sulfur is used in the *vulcanization* of rubber. It was discovered accidently by Charles Goodyear (yes, the tire guy) to convert rubber from a soft gooey material into something that is harder and more useable.

Today, sulfuric acid is the most widely produced chemical in the world, used in industry and research labs.

From the Earth to the moon

Selenium gets its name from the Greek *selene,* meaning "moon." It was discovered 34 years after chemically similar tellurium, the name coming from the Latin *tellus* meaning "earth." Both of these elements are semi-conductors (being between the nonmetals of both oxygen and sulfur and the metal polonium) and are known for having horrible smells: Selenium's smell resembles horseradish, and tellurium has a strong garlic smell.

Both mettaloids are used in alloys, making steel more machinable. Additionally, selenium has replaced lead for most brass alloys that hold liquids due to fears of toxic lead leaching out into people's drinks.

Although alloys are the primary use of these metals, there has been growing interest recently to utilize the semimetals' semiconducting properties, especially for use in solar cell technology. Tellurium, when mixed with cadmium (Cd, in between zinc and mercury on the periodic table), in a thin film only a few micrometers (or 0.001mm) in depth, can make a device capable of converting light energy into electric energy due to the photovoltaic effect. In a nutshell, the photovoltaic effect simply means that light can hit the surface of the material, exciting the material's valence electrons into a higher energy region where they are able to conduct electricity (electricity, after all, is just moving electrons around). Selenium is also used in another thin-film solar-cell material known as copper indium gallium selenide (CIGS).

Marco — polonium!

Discovered in the late 1800s by Nobel-prize-winning Marie Curie and her husband Pierre while working with pitchblende, a uranium-rich ore, polonium is named after Marie's homeland of Poland. Exposure to polonium would cause leukemia and ultimately the death of the couple's daughter, Irène Joliot-Curie. All isotopes of polonium are radioactive and are highly toxic to humans. One isotope of polonium, polonium-210, is speculated to be the poison used to kill Alexander Litvinenko, a former Soviet KGB agent (the Soviet Union's security agency). This same polonium-210 can be found in tobacco leaves, which many studies cite as a potential cause of resulting lung cancer in smokers.

The same isotope was used as the initiator of the Fat Man bomb used on Nagasaki in World War II. Sometimes called an urchin, the initiator is effectively a nuclear weapon's detonator: A nuclear weapon involves a chain reaction that releases neutrons, so a source of neutrons is needed to start the reaction. This urchin was in the middle of the plutonium that ultimately was the fuel source for the Fat Man.

(Re)Active Singles: The Group 17 Halogens

Group 17, also called the halogen group, is one of the most well-understood element groups in the periodic table. The many different structures associated with these elements have given rise to a large body of knowledge about their properties. In fact, the fundamentals of inorganic chemistry are often taught based on the trends in this group.

Almost all elements bond with halogens (these compounds are called *halides*), which means periodic trends can be highlighted according to properties exhibited by elements in the halide form.

Halogens are found easily in nature, but always in compounds and not as individual atoms. In pure form, they exist as diatomic molecules. More often they are found as halides because they form highly reactive diatomic molecules. The chemistry of the halogens is dominated by singly bonded atoms or singly charged anions. They are also known for their small size; additionally, as you go down the halogen group, there's a decrease in the elements' electronegativity and an increase in their metallic character. The halogens exhibit a range of reactivities, similar to all of the other main group elements. They range from the most electronegative and most reactive element in the periodic table — such as fluorine — to some of the least reactive — such as iodine and astatine.

Nearly all the other elements form thermodynamically stable halides based on a typical stability, with higher stability for F > Cl > Br > I. In covalent compounds, this is represented by bond strengths, and in ionic compounds it is represented by *lattice energies* (described in Chapter 8). Halides are common, but when they are with nonmetals they are molecular halides, and when they are bonded to metals they form as ionic halides. Some of the halides are good Lewis acids.

Halogen oxides are typically not very stable, and some oxygen compounds form as *oxoacids* instead (with the exception of fluorine; there are no oxoacids of fluorine).

Some *interhalogen* compounds form as well, and they form an extensive array of both neutral and cationic compounds with each other. These can be diatomic (ClF, BrF, IF, BrCl, ICl, and so on), tetratomic (ClF_3, BrF_3, IF_3, and so on), hexatomic (ClF_5, BrF_5, IF_5, and so on), and even octatomic in the case of iodine heptaflouride (IF_7).

Halide *radicals,* or atoms with unfilled, open shells, are corrosive because they want to strip electrons from other atoms around them to quickly fill up their own empty shells. For example, ultraviolet light from the Sun activates the various forms of chlorofluorocarbon (CFC) at high altitudes, the light gives enough energy to split up the CFC into its constituent atoms, and then chlorine strips electrons from the nearby oxygen atoms. This eventually leads to a depletion in the presence of ozone in the atmosphere, commonly referred to as the ozone layer. Halogen radicals are also corrosive to metals.

Fluorine (F) is one of the most reactive elements and forms chemical bonds with almost all the other elements (with the exception of helium and neon), though it takes higher temperatures to react with noble metals, such as gold, platinum, and palladium. Fluorine can even react with an inert gas (krypton)!

Fluorine was named by French scientist André Ampère in 1812, but it wasn't isolated until 1886 by Henri Moissan. Moissan was able to achieve this by applying electrolysis of potassium bifluoride (KHF_2) in a solution of hydrofluoric acid (HF). This won him the Nobel Prize in 1906.

Fluorine has the strongest electronegativity and is the strongest oxidizing element on the whole periodic table. Fluorine is found naturally in the mineral fluorspar (CaF_2) and the mineral fluorapatite ($Ca_5(PO_4)_2F$). Fluorine is chiefly prepared by electrolysis of potassium fluoride that is first dissolved in anhydrous hydrogen fluoride.

Fluorine can replace Ca^{2+} when in the form of hydrofluoric acid (HF), which is a weak acid with a tenacious capacity for destruction. On the other hand, fluorine fills gaps made by tooth decaying bacteria, and for this reason has been added as fluoride (F^-) to public drinking water in the United States in an effort to improve oral hygiene. It's also used in nonstick cooking pans in the form of poly(tetrafluoroethene), also known as PTFE or Teflon.

Cleaning up with chlorine

Chlorine (Cl) is found all over the planet in seawater, but today it is more often recovered from large deposits underground that formed when ancient seas evaporated, leaving the chlorine behind in the form of NaCl.

In the form of hydrochloric acid (HCl), chlorine was known to alchemists as *acidum salis* or *spiritus salis*. Early chemists called it *muriatic acid,* from the Latin *muriaticus* for pickled in brine because the acid tastes salty, well into the 20th century, and it can still be found in hardware stores under this name today. Pure chlorine, however, was not isolated as a gaseous element until 1774 by Swedish scientist Carl Wilhelm Scheele. Sir Humphrey Davy also isolated it and presented it as a single element when it finally accepted as an element.

Chlorine reacts with many elements, typically rendering compounds that we commonly refer to as salts. Elemental chlorine is obtained by electrolysis of an aqueous salt solution.

HCl is found in the human stomach; it's used to help digest materials as part of the digestive system. Chlorine is also found in many household products, particularly ones that are used for cleaning such as bleach. Prior to the discovery that bleach could whiten clothes, they were laid out in the sun on benches; the name *bleachers* for the seats at ball games is derived from this origin.

Briny bromine

Similar to chlorine, bromide (Br) is found in seawater, but in much smaller quantities (30 ppm) where it is reacted with chlorine to form elemental bromine as illustrated in the equation

$$Cl_2 + 2\,Br \rightarrow 2\,Cl^- + Br_2$$

Bromine was discovered by a German student named Carl Löwig. Löwig produced a sample for his professor that he collected from a spring in his home town. A year later, Antoine-Jérôme Balard isolated bromine from seawater and presented it to the French Academy in 1826.

Because it's found in seawater, marine animals developed techniques for converting it to other forms; for example, organobromides (compounds with carbon and bromine) are made by sponges, corals, seaweed, and even some mammals. In the lab, organobromides are synthesized by scientists, perhaps most notably with the flame retardant tetrabromobisphenol-A (TBBPA). Many organobromides have the ability to hinder combustion, making them excellent flame retardants; however, there's still concern over their role as an environmental pollutant. Other uses of synthetic organobromides on an industrial scale include it being used as a fumigant, because it evaporates easily and leaves little residue behind. However, in the atmosphere it depletes the ozone layer. As a consequence of this the use of organobromides is regulated under the 1995 Montreal Protocol.

The use of bromine can be traced back at least to the Roman Empire, whereby the purple togas that were worn by Roman emperors was dyed with bromine that was extracted from sea snails.

Iodine

Iodine (I) was first isolated and verified as an element in France during the rule of Napoleon. Sir Humphrey Davy visited France to confirm the discovery with his makeshift traveling laboratory. One of the original chemists to claim it as a new element, Joseph Gay-Lussac, coined the name *iode* to represent the Greek word *ioeides* meaning violet-colored, based off of the color of the element.

Iodine is a nonmetal, but it has some metallic characteristics such as its lustrous appearance. When under high pressure, it exhibits electrical conductivity. Iodine is found naturally as sodium iodate ($NaIO_3$); it's then isolated by reduction. Alternatively, it's also found in salt water, and just like bromine, it can be purified by the oxidation reaction with chlorine. In your body, iodine is stored in the thyroid gland.

Iodine has as many as 37 isotopes, many of which happen to be radioactive. Further to this, many of these isotopes form in the event of a nuclear explosion, and should they be released into the atmosphere it's advisable to take iodine supplements to prevent the natural iodine in the thyroid gland from being replaced by a radioactive type. This helps mitigate the risk of cancer forming in the thyroid gland.

Rarely astatine

Astatine derives its name from the Greek word *astatos* meaning "unstable," as it was thought for many years that it would never be found in nature. If you were to gather all the astatine in the Earth's crust together, you would barely have two tablespoons' worth of material. This is because all the astatine in the Earth's crust is only temporarily there, until it decays into another element. It was considered the rarest naturally occuring element for many years until trace amounts of berkelium were discovered, making astatine the second rarest element.

Astatine's most stable isotope has a half-life (the time for half of the material to radioactively decay into another element) of only 12 hours. If you made a gram of it today, you would have only 0.25 grams left tomorrow. This makes astatine difficult to handle, but there's an upside to such a short half-life: It changes very quickly. Some researchers are trying to employ this property for use in anti-cancer research. Astatine-211 gives off a powerful alpha particle (two protons and two neutrons) when it decays into a relatively stable (30-year half-life) bismuth atom. Scientists and doctors hope to harness that power to kill a single cell, such as a harmful cancerous cell, while avoiding large doses of weaker radioactive materials, like current radiotherapies used for cancer treatment that also kill healthy cells.

Lights of New York: The Group 18 Noble Gases

The Group 18 elements are called the noble gases because of their low reactivity. Like a noble man (as opposed to a nobleman) who does not interfere with others, these elements rarely involve themselves in reactions with other atoms. These gases include: helium (He), neon (Ne), argon (Ar), krypton (Kr), xenon (Xe), and radon (Rn).

The characteristic low reactivity of the noble gases is a result of their electron configuration. All orbitals are filled, with no room for more electrons and very high ionization energies are required to gain or lose an electron from a filled configuration. This characteristic helps to define the edge of the periodic table; although they're boring in terms of their reactivity, their very presence confirms the logical organization of the periodic table by providing a stopping point — a point of energy minimum.

The noble gases used to be referred to as the *rare gases* or the *inert gases,* but these definitions turned out to be not as accurate as originally thought.

Argon (Ar) gas, for example, is over 30 times more abundant than carbon dioxide and, therefore, not rare. And xenon is not inert; it's first compounds were created in 1962. When xenon (Xe) forms binary fluorides and oxides as well as fluoride complexes and oxoanions, the stability of these compounds is very low. It's reactivity is related to increasing atomic size as you go down the table, which leads to a decrease in the first ionization potentials. Xenon tetraflouride (XeF_4) is made by mixing one part xenon gas to three parts fluorine gas in a container at 400 °C. Compounds have been confirmed for argon (HArF), krypton (KrF_2), xenon (numerous fluorides, oxyfluorides, and oxides), and radon (RnF_2). It's believed that compounds exist with helium and neon as well, though none have been experimentally proven to date.

Despite xenon's relatively high reactivity (which isn't really high at all) the Group 18 elements are considered *unreactive monatomic gases.* The only other noble gas to form any compound is krypton, which can form a very unstable difluoride binary compound. Compounds of noble gases, typically that of xenon, have oxidation states of +2, +4, +6, and +8.

Argon is the most abundant noble gas; it comprises about 0.93 percent of the atmosphere. Helium, neon, and krypton (Kr) can all be found in air as well, and they can be isolated using fractional distillation of liquid air, except for helium.

In the laboratory, these gases are important because they are inert, so they won't react with chemicals that react with air and/or moisture. Some chemicals ignite just opening the bottle in regular atmosphere, but are important for chemists. To use these chemicals, known as pyrophoric chemicals, scientists handle them in boxes that have been flushed with inert gases, commonly argon because it's relatively affordable and heavier than air.

Helium (He) is produced by cryogenically distilling natural gas. Helium has a low boiling point and will liquefy only under high pressure. Liquid helium, therefore, is used to cool environments to very low temperatures. This can be important for scientific applications where temperatures close to absolute zero are required; helium's freezing point is only 4 °C warmer than absolute zero. For comparison, water freezes at 273.15 °C warmer than absolute zero.

Helium comes from the emission of alpha particles during radioactive decay, some of these radioactive particles from all the way from the sun to make helium. (Next time you hold a floating party balloon just remember how far that helium traveled all the way from the Sun!) Radon, on the other hand, is formed by the decay of radium and uranium and all isotopes of radon are radioactive.

Xenon does form some binary fluorides and oxides as well as a few fluoride complexes and oxoanions, but the stability of these compounds is very low and are reactive compounds. The only other noble gas to form any compound is krypton, which can form an unstable difluoride binary compound.

Of course we can't conclude the noble gases without talking about neon lights. In short, a glass tube is pumped down to very low pressure and filled ever so slightly with a small amount of noble gas — usually the pressure inside a neon light's glass tube is about a hundredth of the pressure outside the tube. This is why neon lights are so fragile and commonly are seen not functioning. A glass blower shapes the tube into whatever the sign is to say. Then, when electricity is added to the tube, the electrons of the noble gases get excited at the center of the tube where the electricity is flowing, but eventually cool down just enough for the electrons to return to the energy level they started with. Each noble gas's electrons have a distinct amount of energy it can absorb to get excited — when they relax, they emit that exact amount of energy, some of which happens to be at a specific energy level that relates to visible light. This is why neon glows blood red, argon glows lavender, helium glows a yellowish-orange, and krypton glows a light blue. For other colors, the light maker coats the inside of the glass tube with an ultraviolet phosphor, or a chemical that emits light of a certain color when ultraviolet light interacts with it. Then, the lamp maker adds a drop of mercury into light. It takes only a tiny amount of that droplet for a mercury gas to emit light. Although you can't see the UV light being emitted by the mercury, it activates the phosphor and allows you to see whatever color the phosphor emits.

Chapter 13

Bridging Two Sides of the Periodic Table: The Transition Metals

*T*he transition metals (or T-metals, as we call them) span the area between the alkali and alkaline Earth elements (on the far left), and the main group elements (on the far right). The T-metals are subdivided into three main categories according to the electronic structure, and what effect the change in electronic structure has on the size, chemistry, and physical properties of the T-metals. These elements are called the transition metals because they occupy an interstitial part of the periodic table that transitions from elements with only s-orbitals (alkali and alkaline elements), to those with both s- and p-orbitals (main group elements). If the periodic table was looked upon as a landscape, the area occupied by the transition metals acts as an isthmus between each side.

In this chapter, we look primarily at the main transition metals, also referred to as the *d-block elements*. (Note that the other transitional metals are covered in Chapter 14.) The elements we present in this chapter are important for industry and chemistry research for use as magnetic materials and catalysts. We start you out with a description of the key characteristics of the d-block elements and then explain how the filling of d-orbitals drives their chemical interactions and bonding with other elements. At the end of this chapter, you find important information about the simple steps for naming transition metal compounds.

Getting to Know Transition Metals

All of the transition metal elements share a few similar attributes such as:

✔ They form hard and strong materials with high melting and boiling points.

✔ They form alloys with one another and with other metallic materials.

✔ Due to unfilled or partially-filled orbitals, they form paramagnetic compounds.

✔ They have low ionization potentials, which means they conduct heat and electricity.

✔ They form lustrous, malleable, and ductile materials.

✔ Most dissolve easily in mineral acids such as hydrochloric acid.

✔ They react with halogens, sulfur, and other nonmetals in the presence of some thermal source.

✔ They have a variety of valencies and various oxidation states.

✔ They are often refractory; for example, silver has been used in mirrors.

✔ The trend in bond strengths for the T-metals is opposite to that which is found for the main group elements; for example, the metal-ligand bond strength is ordered by 3d << 4d < 5d.

Sorting T-metals into series

The transition metals can be separated into three *series* based on the filling of d-orbitals and coinciding with their row on the periodic table. The *first transition series* of t-metals includes scandium to copper and zinc (zinc behaves differently to other first transition series elements because it doesn't have partially filled 3d-orbitals and is never ionized).

The *second transition series* from yttrium to silver all have a partially filled 4d-orbital. This includes cadmium, but similarly to zinc, cadmium has no partially filled d-orbitals; therefore, it is not technically considered among the T-metals. All the second transition series elements except for yttrium form important compounds such as the Wilkinson catalyst - rhodium(I)tris-(triphenylphosphine)chloride, $\left[Rh(Cl) \, (PPh_3)_3 \right]$.

The third transition series are 5d-orbital elements that span from lanthanum to gold. Lanthanum is considered an analog of the elements known as lanthanides; these are not T-metals and don't react similarly to T-metals, thus they are considered separately (see Chapter 14).

The radii of the first series of T-metals are the smallest of the various types. The heavier elements both as pure metals and as ions tend to be much larger than that of the first series elements. The sizes of the radii, going from smallest to largest, is 3d < 4d < 5d ≤ 6d. The 4d and 5d elements are similar in size due to a condition known as *lanthanide contraction.* (You can find details on this in Chapter 14.) Radii in general decrease in all series as you go from left to right on the table.

Separating T-metals from the main group

There are few distinct ways that transition metals distinguish themselves from main group elements. For example, the ionization energies of the T-metals are higher than that of the main group elements, and this results in high melting and boiling points when compared to the main group elements.

The transition metals often have a variety of oxidation states and usually have an oxidation state that's less than that of their group number. The oxidation state of the main group elements, however, is usually the same as the group number. For example, the main group elements Na^+, Mg^{2+}, and Al^{3+} all have a single oxidation state that's the same as their respective group numbers.

Take a moment to look at the periodic table as a large system that spans from one side to the other, where each piece is somehow related to every other piece. Notice that on the far left of the table, the alkali and alkaline Earth metals are reactive atoms and undergo vigorous reactions with water. Take scandium, for example. This is on the left side of the transition metals section, is a reactive element, and is very energetic. But on the far right side of the transition metals, there are atoms — such as copper or gold — that aren't very reactive at all. And beside them, in the main group metals, you find tin. Tin is so inert that it was used to line the inside of metal cans (tin cans) — that is, until it was found that it is partially soluble in water, but that's another story. So you see how the transition metals fit in the periodic table between Groups 1 and 2 on the left side, and the rest of the main group elements on the right side.

Partially Filling d-Orbitals

The chemical and physical properties of transition metals are based on the continuous filling of d-orbitals with successive electrons. In Chapter 2, we explained how atoms are structured with different types of electron orbitals. In brief, we showed that when adding more and more electrons to the nucleus, the electrons like to sit at a certain distance from each other and from the nucleus. For this reason, there are some differences in chemical reactivity of the 3d elements compared to both the 4d and 5d series elements due to the differences in the d-orbital shapes and sizes. This section focuses specifically on the 3d series and leaves the details of 4d and 5d series for Chapter 14.

The first transition series or 3d-orbital series has a weak metallic character and less of a tendency for covalent bonding. This series also has strong oxidizing properties when in higher oxidation states. It forms many compounds with unpaired electrons, making it useful for catalysis and as magnetic materials.

Calculating an effective nuclear charge

The effective nuclear charge (Z^*) is the net charge felt by an electron when it is in an atom. It's affected by the full nuclear charge (Z) of the atom and the shielding constant (S). It can be calculated simply as $Z^* = Z - S$.

The full nuclear charge (Z) comes from the protons and neutrons inside the nucleus, but the shielding constant (S) comes from presence of the other electrons in the atom. The shielding constant, or screening constant, is also referred to as the Slater shielding constant after J. C. Slater who developed the concept in 1930.

In multi-electron atoms, such as the transition metals, the energies of the electrons in the orbitals depends greatly on the energies of all the other electrons that are present. s-, p-, d-, f-, and g-orbitals experience a split in energies and so the subshells drop to lower energies. Therefore, as you move across the table, the energy levels go up and down in a manner, that at first might appear as random. This is due to the effective nuclear charge.

As more electrons are added to the atoms as you continue to go from left to right on the periodic table, the effect of the screening power varies. For the first 18 electrons, the screening power varies because of the electrons' inconstant penetration into the nuclear core. For example, 3d electrons barely penetrates the core, but the 4s and 4p electrons penetrate more deeply.

The same sequence of energy transitions occurs for both the 3d and 4d-orbitals. This unique characteristic illustrates why the T-metals are positioned in the middle of the periodic table: the energy levels at the start of the series are similar to that of the alkaline earth metals (on the left of the periodic table), then increases in the middle, but then decrease on the right side to values found in the main group elements. However, in the middle, they're quite adaptive due to the possibility of various oxidation states they can adopt.

Forming more than one oxidation state

Most transition metals have more than one oxidation state. The oxidation state of a T-metal can determine the color of a compound that it makes. It was in observing this that the field of organometallic chemistry began to flourish. It takes only small changes in the metal center and the coordination compound to make great changes in colors; for example, two arsenic compounds that have been known since antiquity and used as pigments in paints are arsenic sulfide (AsS) and arsenic trisulfide (As_2S_3). They're known commonly as Realger (red color) and Kings Yellow (yellow color), respectively.

Removing electrons from T-metals become more difficult moving from left to right in the series. The first T-metal elements most often have a relatively high oxidation state, whereas the later elements tend to exhibit low oxidation states. Figure 13-1 illustrates the many oxidation states of first row transition metals.

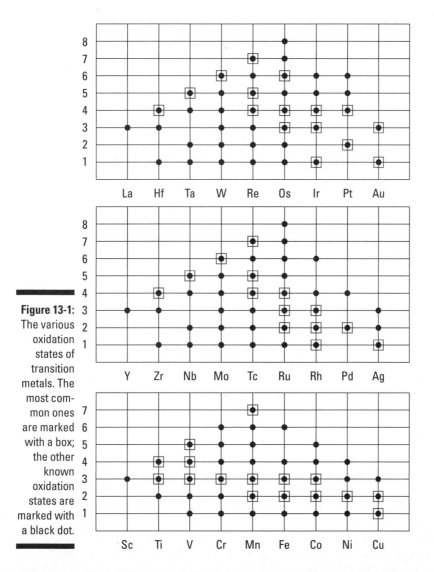

Figure 13-1: The various oxidation states of transition metals. The most common ones are marked with a box; the other known oxidation states are marked with a black dot.

The first transition series, with the exception of scandium, tend to form 2+ cations. To do this, they must lose two 4s electrons from the valence shell. The 3d and 4s orbitals have similar energies so it's possible to lose a 3d electron too, giving rise to a cation with a 3+ oxidation state. A third electron can be removed also, but it requires much higher energy to do so. More electrons can be lost, too, and this is done by the sharing or loss of additional d-orbital electrons.

The second and third transition series have trends in oxidation states similar to the first series. For higher oxidation states, the stability tends to increase as you go down the periodic table. For example, rust is quite stable (Fe_2O_3), but osmium tetraoxide (OsO_4) is a volatile compound. It's so volatile that it can pass through the membrane material of an eye and bind to the back of the eye, causing blindness.

Splitting the Difference: Crystal Field Theory and Transition Metal Complexes

Transition metal complexes have always amazed chemists because of the interesting colors that can be made. The different colors aren't explained by differences in geometry that are described by valence bond theory. Rather, chemists turn to a bonding model called *crystal field theory* to understand why transition metal complexes exhibit a dramatic range of colors.

In crystal field theory, you assume that metal-ligand interaction occurs solely through Coulomb forces — that is, they are electrostatic in nature and are driven by charge balance through electron sharing in the metal-ligand system. This can happen in two different ways: Either there is some ion-ion interaction or there is an ion-dipole interaction.

For the case of ion-ion interactions, imagine two spherical and oppositely charged ions such as $Co^{3+} + Cl^- = CoCl_3$. They interact, share electrons, and form a compound that is of net neutral charge.

The interaction can also occur between an ion with a polar ligand. Take the example of NH_3 that has a positive dipole δ^+ (H) and negative dipole δ^- (N). As such, the bonding occurs from the electropositive metal center to the electronegative part of the ligand; remember, opposites attract. In practice, this means the nitrogen points to the metal center, such as the case with $Co^{3+} + (NH_3)_6 = [Co(NH_3)_6]^{3+}$. The exact way that the metal and ligands bond depend on the interaction of the orbitals between the metal and the ligand. There are three types of interaction between the metal and the ligand: σ overlap, π overlap, and dπ-pπ (back) bonding.

Crystal field theory is used to elaborate the effect that ligands have on metals and how they interact with each other. Ligands interact with metal centers according to the position of the d orbital relative to the ligand field. If the lobe of the orbital is pointed in the direction of the ligand, it experiences a repulsive force, raising the energy level in that orbital. This is a secondary effect of the metal-ligand interaction and accounts for only 10 percent of the total metal-ligand interaction energy, but it has a profound effect on the electronic, magnetic, and stability properties of metal complexes.

Crystal field theory has since been surpassed by ligand field theory because it takes into greater account the presence of covalent bonding. Nonetheless, crystal field theory is still a powerful theory, although simple, as it explains magnetism, electronic spectra and binding strengths of transition metal complexes.

Dividing d-orbitals

To start this discussion, become familiar with the shapes and symmetry of the d-orbitals. These are shown in Figure 13-2. Notice how there are five orbitals and they're grouped into one group of two and one group of three. They're grouped according to symmetry elements (you can get a recap of symmetry in Chapter 7). In the figure, t refers to triply degenerate (same energy), e refers to doubly degenerate, and g refers to being symmetrical with respect to the center of symmetry.

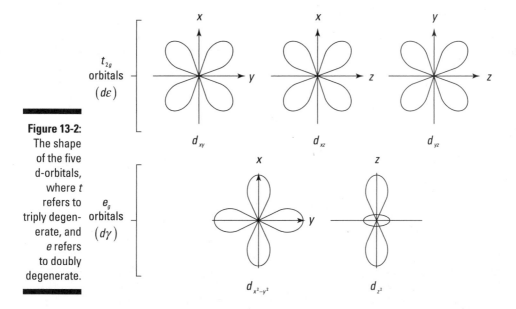

Figure 13-2: The shape of the five d-orbitals, where *t* refers to triply degenerate, and *e* refers to doubly degenerate.

Now take the example of a free floating metal atom. Imagine it's just free floating in space and there's nothing near to influence it in any way. In this situation, the electrons (all of like charges) repel each other equally and so they gladly occupy any orbital that is free and available. In this case, all the orbitals have the same energy, or energy level. This is known as *degeneracy;* in isolated atoms, all the d-orbitals have the same energy so they are degenerate. In short, this is like the ground level; we start with this assumption and add on to it.

The next thing we add to this model is the presence of ligands. As a ligand gets closer and closer to the free floating metal atom, the electrons in the d-orbitals start to repel the electrons of the ligands (like forces repel). This makes the electrons in the metal want to move to whichever orbital is furthest away from the ligands. And the position that the ligands take up around the metal atoms depend on the coordination number of the atom. To be more precise, it depends on whether it's an octahedral or a tetrahedral complex.

Remember we said earlier in this chapter that transition metals have partially filled d-orbitals? Well, this becomes very important now because the electrons are going to want to go somewhere to get away from the ligands. The ligands act like a sort of force field (it is called crystal *field* theory after all). When the ligands come close to the metal (that they are attracted to), the electrons around the metal can feel the force field of the ligand. So the electrons in the d-orbitals try to get away from it, and they try to go to the place (orbital) that requires the lowest amount of energy. So, which orbital would that be? you might ask; this is where it gets interesting.

Remember, there are five d-orbitals that are grouped according to symmetry, and each orbital can hold up to two electrons. All the orbitals that have the same symmetry are expected to have the same energy level (degenerate), so they are placed alongside each other. The orbitals are represented as boxes that can be filled with arrows, that represent the electron and the spin. The electrons can either be spin up with an arrow pointing up, or spin down with an arrow pointing down, this follows from the Pauli exclusion principle (see Chapter 2).

As you can see from Figure 13-3, the energy levels get split according the symmetry of the orbital, one of higher energy than the other. The energy difference between the two energy levels is called the field splitting parameter, Δ_0, where the O represents the fact that it is in the octahedral field. If the value of Δ_0 is small, it's considered a weak field ligand; and if the value is large, it's considered a strong field ligand. The magnitude of the field splitting affects the spectral properties of the complex.

In practice, electrons prefer to occupy an unoccupied orbital first, before sharing an orbital with another electron. If they do share an orbital, then they must do so with an opposite spin. If there's a case where there are lots of unshared electrons, the complex is said to be a *high spin complex,* but if the electrons are all paired together, then it's called a *low spin complex*. This affects the magnetic properties of the complex.

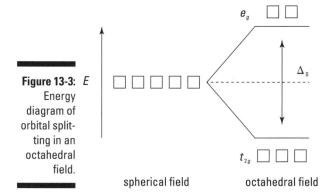

Figure 13-3: Energy diagram of orbital splitting in an octahedral field.

Figure 13-4 shows an example of both a low-spin, strong-field ligand and a high-spin, weak-field ligand. This is demonstrated for a metal with four electrons in the d orbital that's more commonly written as d^4.

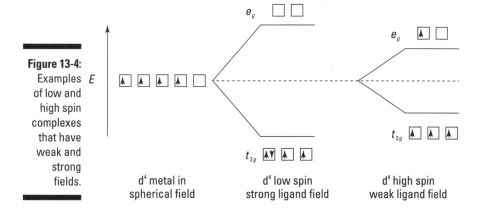

Figure 13-4: Examples of low and high spin complexes that have weak and strong fields.

Absorbing light waves: Color

As we mentioned many times, transition metal complexes very often have quite brilliant colors, and the same metal center can exhibit many different colors. The color of the compound is determined as the magnitude (size) of the field splitting parameter, Δ_O.

When light hits the complex, some of the energy is absorbed by the transition metal complex, and this is then observed by someone looking at the complex. For example, take $[Ti(H_2O)]^{3+}$, a hydrated Ti^{3+} ion, this has a reddish-violet color. This occurs because the yellow and green part of the light is absorbed by the complex, the energy in the light is used up to excite the electron. And so the transmitted (observed) color is complementary to the absorbed color.

The size of Δ_O affects the optical properties of the complex and can be measured using UV-Vis spectroscopy (see Chapter 22 for more details). This is the easiest way to determine the value of Δ_O.

The magnitude of Δ_O is dependent on three factors such as the nature of the ligands, the charge on the metal ion (oxidation state), and from what row the transition metal comes from (first, second, or third).

The energy of Δ_O is of the order of chemical bond energy, it increases with increasing oxidation number, and it increases going down a group.

The relationship to the ligands depends on whether it is a strong field ligand or a weak field ligand. There is a spectrochemical series that orders the ligands according to how they change the color of the transition metal complexes. This series is proportional to ligands, regardless of that metal center in use. It's almost as though the ligands are like tuning forks; they set the frequency and the metal atom produces the colors to match that frequency.

For example, there are at least three different cobalt(III) complexes that can be isolated when $CoCl_2$ is dissolved in aqueous ammonia, and the product is then oxidized in air so it can form into the 3+ oxidation state. Table 13-1 lists the formulas of some cobalt complexes and the colors they display.

Table 13-1	Cobalt Complex Formulas and Colors
Formula	*Color*
$[Co(NH_3)_6]Cl_3$	orange-yellow
$[Co(NH_3)_5(H_2O)]Cl_3$	red
cis-$[Co(NH_3)_4Cl_2]Cl$	purple
trans-$[Co(NH_3)_4Cl_2]Cl$	green

In short, all of these complexes have different colors simply by virtue of the number of ammonia ligands that are complexed to the metal center, and the resulting shapes of the isomers that are created from them.

Building attraction: Magnetism

Magnetic behavior in materials arises due to the presence of unpaired electrons. Just as the metals can exhibit wildly different optical properties due to differences in crystal field splitting, so too can the magnetic properties.

A Gouy balance is a simple tool that can be used to measure the magnetism of a material. There are two ways in which a complex responds in a magnetic field — it's either attracted to it, or it's repelled from it. When it's attracted to the magnet it's referred to as *paramagnetic,* and when it's repelled it's known as *diamagnetic.* The different response can be seen on a weighing scale because a magnetic material might get pulled more onto the scales, for example, making it seem heavier. How much heavier it weighs gives an indication of how strongly it's effected by the magnetic field. The more unpaired electrons there are, the heavier it seems. However, if it's repelled by the magnetic field, the material becomes lighter. By this way, both the direction and the magnitude of magnetism can be measured.

Materials called *loadstones* are possibly the oldest known magnets. These were made of magnetite, which is an iron-containing mineral. Many other transition metals are used for their magnetic properties, but iron is a classic example of a known magnetic material.

Magnetism is an important property; for example, it's used to determine the direction to the north and south poles. The Earth has an inherent magnetic field. When using a compass, seafarers and other travelers and tradesmen could navigate vast distances thanks to this understanding. A rudimentary compass can be fashioned using a magnetized pin (or a light piece of metal wire) that floats on water using a leaf or a piece of silk. The Earth's magnetic field causes the pin to line up along the north-south axis as long as the water is still and doesn't interfere with the magnetism. To magnetize a metal pin, you can rub it against some other magnetic material. In fact, this is how the first magnets were made — by making intimate contact with a loadstone.

Consider $[Co(NH_3)_6]^{3+}$ and $[Co(NH_3)_6]^{2+}$ are two similar compounds on paper, but in reality they show some marked differences in their behavior in a magnetic field. The d-orbitals of the central cobalt atom undergo a splitting due to the bonding of the six ammonia ligands. But the degree to which they split is markedly different among each one.

The compound with the Co^{2+} (with three unpaired electrons) undergoes only a small d-orbital splitting. But the compound with the Co^{3+} (four unpaired electrons) undergoes a much larger d-orbital splitting. This material is paramagnetic, as opposed to the prior compound that is, in fact, diamagnetic instead.

Recent advances using nanoparticles of iron oxide and other transition metals have been touted as a means for targeted drug delivery within the body. This is because, as a nanoparticle, they have the capacity to have all the dipoles of the atoms line up and act in unison, making them extremely sensitive to magnets. This happens when the atom becomes magnetized which makes it polarized. It then has a magnetic pole (just like the north and south pole of Earth). All of the poles of the atoms can align with each other collectively, thus increasing the total magnetic character of the nanomaterial. They can be put in the place where they are required, either to deliver specific drugs or to act as contrast agents to improve imaging quality for example.

Electronic Structure and Bonding

In understanding the profile of the T-metals, consider simultaneously a few interrelated properties. In the first instance, consider the electronic structure and the type of bonding they undergo as you go along the table.

Going from left to right on the table, the electronegatvity of the elements tends to increase. This means that as you from left to right, as with the rest of the periodic table, the reactivity tends to increase for metal atoms.

But going down vertically, the elements become less reactive and so are often less used in industry or research for chemical reactions. The 4d and 5d elements are relatively inert in comparison to the 3d elements.

There is an important difference in electronic structure between the three classes of T-metals. The d-block elements that we focused on in this chapter have larger orbitals than the s, p, or f orbitals of the lanthanide and actinide elements (see Chapter 14 for details on these).

When electrons are further from the core, and when they are not hindered by the presence of other electrons in nearby orbitals that either raise their energy levels or screen them altogether, they more easily undergo interactions with nearby atoms and are more easily affected by their environment.

The electronic structure shows somewhat erratic variation in energies as you look across the transition metal elements. Keep in mind how the d-orbitals span physically so far out from the nucleus while penetrating into the nucleus simultaneously. This affects the bonding of the elements. The difference in energies causes some of the d-orbitals to split into lower electron levels, and this in turn affects the symmetry of the molecules formed.

High symmetry molecules exhibit only slight covalent bonding character in the bonding between a metal center and an adjoining ligand, for example. Covalent character is based on directional bonding, and high-symmetry molecules can often not make directional bonds to ligands so it's more electrostatic or ionic in nature.

Reacting with other elements

The reactivity of T-metals to other non T-metal elements varies widely. They are largely found in nature bonded with oxygen as oxides. The reactivity toward bonding with oxygen tends to decrease moving left to right along the table, but there are some abnormalities to this rule.

Ligand field theory

Ligand field theory is used to teach bonding arrangements of the transition metals with ligands. It's a theory that's more complicated than crystal field theory because it accounts for covalent bonding in organometallic compounds whereas crystal field theory treats the metal-ligand bonds as purely ionic. Start by counting the electrons in the d-orbital, regardless of whether or not the material is ionic. If the material is covalently bound, you can use the 18-electron rule to count the number of electrons (including the d and s orbital electrons) and determine the structure of the materials (see Chapter 15). This is also true when the metals are in a low oxidation state. This is easily determined after you know the group number of the metal. For example, iron is in the Group 8, and so you consider it to have eight electrons.

Transition metal oxides are used in various ways. For example, chromium is often used to coat materials because the oxide is resistant to wear and tear due to the presence of a protective oxygen layer on the surface. Another example is titanium. It is often used in climbing materials that are designed for extreme weather conditions that will generate a lot of wear and tear. The titanium dioxide layer is more resistant to corrosion and degradation than the titanium itself. This effectively means when the TiO_2 layer is removed, it heals itself by oxidizing rapidly and so maintains the strength of the titanium metal and resistants corrosion at the same time. In some cases, however, the oxides are more brittle than the pure metal. For example, rust (iron oxide) is more brittle than pure iron.

The transition metals form carbides, nitrides, and borides with nonstoichiometric ratios. Some of these materials are *interstitial compounds,* which means there are large gaps in between the metal atoms that can be filled by other small elements such as hydrogen, nitrogen, and carbon. When interstitial compounds form, the additional atoms are locked into place in the metal framework, making these materials stronger than the metal by itself. This is important in a number of industrial applications such impregnating a small amount of carbon into an iron matrix in order to make steel.

Interstitial compounds are one way in which the hydrogen economy can be realized because the hydrogen can both get into and back out of the gaps of some T-metal compounds quite readily. These materials used in the hydrogen economy are known as metal hydrides.

T-metals also have a tendency to form compounds in the solid state, whereby the tendency to form cationic compounds decreases from left to right along the series.

Binding energies initially increase, peak in the middle, and then decrease at the end, making somewhat inert elements known as coinage metals such as copper, silver, and gold. And finally, metal-metal bonding changes between series. The first-row elements prefer metal-carbonyl bonding, but the heavier elements in series two and three are more prone to metal-metal bonding.

Creating coordination complexes

A coordination compound is a compound that has a metal atom at its center, surrounded by a group of molecules attached by coordinate covalent chemistry.

Cisplatin is one such coordination complex. The metal at the center is platinum, and it has two ligands of NH_3, and two ligands of Cl^-. These are called the ligands, and the atom in that ligand that is directly attached to the metal center is called the *donor atom*. In this case, the donor atoms are nitrogen and chlorine.

Making a coordination compound is based on a Lewis acid-base interaction (see Chapter 5). The ligands act as a Lewis base (electron pair donor) with the central metal atom acting as a Lewis acid (electron pair accepter).

The electropositivity of the metal center determines the number and type of donor atoms that can bind to it. And the size and electronic configuration of the metal center determines the placement or arrangement of the ligands about the central core. This is due to the number of d-orbitals that are occupied or available for bonding, which is affected by the electronic structure of the T-metal.

The coordination number of the metal center depends on the size of the metal ion, the charge of the ion, the electronic configuration, and on the size and shape of the ligands. Coordination numbers of 4 and 6 are most common, but they can range from anywhere between 2 to 9.

Adsorbing gas: T-metals in catalysis

Transition metals are commonly used for catalysis because their large d-orbital size causes increased reactivity. In particular, they have a knack for adsorbing gases. They can undergo catalytic reactions of the gases, ejecting the products when complete (see Chapter 16). For example, this is how the catalytic converter of a car functions. A palladium catalyst breaks down the gas from the exhaust of the engine and then converts the toxic gas to something less toxic.

The exhaust of an engine has both benign and harmful compounds in the emission; it's desirable to convert the harmful compounds to something safer. These compounds include carbon monoxide (CO), volatile organic compounds (VOCs), and nitric oxide (NO) and nitrogen dioxide (NO_2), which are nitrogen compounds collectively known as NOx (pronounced as nox). In the reduction stage of the catalyst, platinum and/or rhodium reduces the nitrogen compounds to create $N_2 + 2\,O_2$. In the oxidation stage, platinum and/or palladium metal catalyze the breakdown of carbon compounds by essentially burning them (oxidizing) in a reaction similar to $2\,CO + O_2 = 2\,CO_2$.

Chapter 14

Finding What Lies Beneath: The Lanthanides and Actinides

*T*his chapter zooms in to focus on a very specific section of the periodic table, where 28 elements crowd around each other in the late stages of the transition metal group. The two rows illustrated across the bottom of the periodic table (see Chapter 2) is where you find the lanthanides and actinides.

The chemistry of these elements is dominated by the filling of f-shell electrons. The unique thing about these elements is that there are no other sections of the periodic table where the effect of f orbitals is noted. Main group elements are dominated by chemistry of s and p orbitals (see Chapter 12), and the transition metals are dominated by chemistry of the s and d orbitals (see Chapter 13).

In this chapter, we describe these elements that are tucked into the last row of the periodic table and some of their unique characteristics. For example, we look at how and why lanthanides undergo lanthanide contraction, how the actinide elements are mostly artificial (man made in laboratories), and how they follow trends similar to those observed in the transition metals.

Spending Quality Time with the Rare Earth Elements: Lanthanides

The *lanthanides* consist of the 14 elements following lanthanum (including lanthanum) with increasing atomic numbers from Z=57 to Z=71. Lanthanum acts as a template for the rest of the elements in this group — they're similar to each other, and chemists use the label *Ln* as a symbol for elements in the lanthanide group. The chemistry of the lanthanides is understood according the partial filling of seven orbitals of the 4f shell.

Lanthanide elements used to be called *the inner transition metals* but are more commonly referred to as the *rare Earth elements*. All but one of the lanthanide elements occur naturally. (Promethium doesn't occur naturally and must be synthesized. It has a half-life of 2.6 years and is radioactive.) Lanthanide elements are often found in minerals such as monazite ($LnPO_4$) and bastentite ($LnCO_3F$). The more reactive ones can be extracted by reduction of $LnCl_3$ with Ca. When these are used together they are known as a *mischmetal,* which means a mixed metal material that is composed of at least one rare Earth element.

Electronic structure

The lanthanide elements differ chemically according to the filling of seven orbitals from the 4f orbital. The electronic configurations span from lanthanum $[Xe]5d^16s^2$ up to lutetium $[Xe]4f^{14}5d^16s^2$. As you move across the series from left to right, the orbitals that experience the greatest addition of electrons are the f orbitals. Here it is broken down such that each line shows the changes in each orbital.

$$5d^1 \rightarrow 5d^1$$
$$4f^1 \rightarrow 4f^{14}$$
$$6s^2 \rightarrow 6s^2$$

Notice that there is no change of the d or s orbitals, only in the f orbitals. This filling of f orbitals ranges from $4f^1$ (for Ce) to $4f^{14}$ (for Lu).

The 4f electrons play only a small role in bonding. These metals are highly electropositive with a +3 oxidation number being typical. Because of this, the 4f electrons are similar in terms of physical and chemical properties. Lanthanide chemistry changes gradually as you move across the series; they are typically +3 oxidation state.

Reactivity

The early lanthanides (on the left side of the periodic table) are more reactive than the late lanthanides because the basicity (the measure of ease in which an atom will lose an electron) decreases from left to right. In other words, lanthanum would be the most reactive in this series because it more readily gives up an electron, and lutetium ise the least reactive because it more likely holds on to its electrons.

The 4f orbitals in general are much less reactive than the 5d orbitals (transition metals). The f orbitals do not span out as far into physical space as the d orbitals, so they are harder to reach and harder to do chemical reactions with. Additionally, the 4d and 5d elements are relatively inert in comparison to the 3d elements. Therefore, not only are the lanthanides less reactive because of the employment of 4f orbitals, but any d orbitals that they might employ are going to be less reactive than the d orbitals of the earlier d block elements.

Oxidation states

The chemistry, applications, and separations of the Ln elements are predominantly affected by the Ln^{3+} oxidation state; it's the most stable state. All the Ln elements form halides of the form LnX_3, and oxides Ln_2O_3. They can also form hydrates readily, and the Ln^{3+} ions are soluble in aqueous liquids.

Oxidation states such as 2+ and 4+ are possible, but they are far less common. One such example includes cerium (Ce^{4+}) that can form oxides with the formula CeO_2. This is known as ceria.

Ionic radii

As you go along the periodic table, the sizes of the elements get smaller. This affects the coordination chemistry, making the later Ln atoms have a lower coordination number typically.

The size of the ionic radii sits in a range from 1.04 Å for La^{3+}, down to 0.86 Å for Lu^{3+}. The large sizes of the Ln^{3+} allows a relatively large amount of space around the nucleus, which in turn allows for high coordination numbers with respect to the various coordination compounds that they can make.

The organometallic chemistry of Ln elements isn't rich with examples because they tend to form ionic bonds rather than covalent bonds. For example, they aren't able to bind strongly with a neutral ligand such as carbonyl (CO), which is one of the most important ligands in organometallic chemistry. (See Chapter 15 for more details on organometallic chemistry.)

Ln^{3+} atoms tend to bond according to ionic bonding properties, but those that have 5f orbitals occupied can have some covalent character that's better described as a *covalent hybrid*.

Melting points

The lanthanides make soft metals. Figure 14-1 illustrates the linear increase in melting temperature for each of the elements from lanthanum to lutetium.

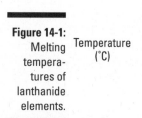

Figure 14-1:
Melting temperatures of lanthanide elements.

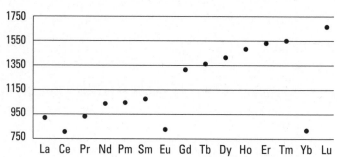

Because they have similar sizes, the alloys of these elements are soft and malleable. Examining the melting temperature is one of the best ways to see differences between the lanthanide atoms.

Lanthanide contraction

Lanthanide contraction explains a situation where the size of the ionic radii for the lanthanide elements are smaller than expected. This result is due to poor shielding from 4f electrons (for more on shielding see Chapter 13). The shielding effect is a phenomenon by which the inner-shell electrons shield the outer-shell electrons so they're not affected by nuclear charge.

In short, the shielding effect for electrons follows the order with decreasing potential going from s > p > d > f. So as the 4f electron shells get filled up in the lanthanide elements, they are less and less able to shield outer (5th and 6th) shell electrons. When the shielding is not as good, the positively charged nucleus has a greater attraction to the electrons, thus decreasing the atomic radius as the atomic number increases.

As a result, the decrease in the total radius is more pronounced for lanthanides than for the other elements as you go from left to right on the periodic table. This trend is illustrated in Figure 14-2.

The effect of lanthanide contraction can be felt for all the other elements that proceed after the lanthanides.

Figure 14-2:
The sizes
of the
lanthanide
elements
showing
the relative
sizes of
the atomic
radius.

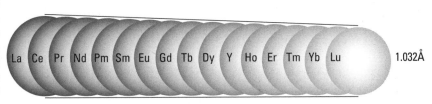

1.032Å

Separating the lanthanide elements

It's been nearly impossible to separate these elements from one another due to the characteristics of their f orbitals and lanthanide contraction. Only recently has some of the chemistry been realized due to advances in separation techniques.

Because it's so difficult to do selective chemistry with a mixed batch of lanthanum atoms (separations based on differences in reactivity is very challenging), they are most often separated first according to their solubility. The key to the separation of lanthanides lies with the slight changes in the solubility of the Ln^{3+} ions. Slight changes in pH in solution can cause the changes in the stability constant for the ions, making it either less likely or more likely to remain in solution, precipitate out of solution, bind to a substrate, or remain in solution to be eluted.

The first popular method for separation was based on how they would stick to a substrate when passed over it, such as a large sieve in a chromatography column. Chromatography is well suited to separating molecules by exploiting slight and subtle differences in the binding affinity with the substrate. The ions that don't bind so well get *eluted* much faster because they wash through the column much easier.

The material that's washed through the chromatography column is called the *eluent*. This is similar to how you might run water through cooked pasta to remove the excess starch, the starch is soluble in the water and is removed from the pasta substrate. This same technique is used to separate molecules because the properties of the solvent that you wash through the material alter the solubility of the materials you're trying to separate.

There are several techniques used to separate the Ln elements, but most of them exploit the difference in solubility and *complexing constants*. The measure of how well the Ln^{3+} cations bind with the complexing molecules is called the *complexing constant;* the value is unique to each of the Ln atoms.

- **Precipitation:** The solubility of Ln^{3+} ions can be altered either by changes in pH (destabilizes the ion according to electrostatic forces in solution) or by changes in polarity (alters the miscibility). The polarity of the solvent affects the solubility of an Ln^{3+} ion, eventually causing it to be immiscible. Coupled with centrifugation, the heavier of the ions can be forced out of solutions and separated using the appropriate centrifugation protocol.

- **Fractional crystallization:** Fractional crystallization can be used to successively crystallize Ln^{3+} ions, whereby the solubility difference for the lanthanide elements can be exploited. In this way you can crystallize a particular fraction of the solution that contains the specific lanthanide element that you seek.

- **Complex formation:** In water, the Ln^{3+} ion is slightly acidic, with a general increase in acidity going from lanthanum to lutetium. This increase in acidity also correlates to the decrease in the ionic radii from left to right. A consequence of this allows separations of Ln^{3+} atoms to occur as they will have a different complexing strength to oxygen donor ligands such as ethylenediaminetetraacetate (EDTA). The earlier lanthanides complex less strongly with the EDTA than the later Ln^{3+} ions, so they can be purified first because they won't stick to the EDTA and don't elute as fast in a chromatography setup.

- **Solvent extraction:** Solvent extraction can also be done using porous materials such as zeolites (see Chapter 18). The porous materials can be made to have a particular binding ligand, whereby the large surface area of the porous materials has a layer added to it with a particular functional group. One such group is a sulfonated polystyrene group, SO_3H.

- **Ion exchange:** The effect of oxidation state can also be considered for some of the Ln atoms because some can change to a Ln^{2+} or Ln^{4+}. This changes the valency of the atom, which in turn alters the reactivity of that atom. In the +2 or +4 oxidation state, the capacity for complexing to other types of ligands can be exploited to selectively remove particular ions.

- **Oxide formation:** Lanthanides can also be separated according to their tendency to form oxides based on an appropriate thermal treatment. The first to react will solidify sooner and can be removed, the least basic turns to oxide first.

Using lanthanides

Compared to the rest of the periodic table, the lanthanide elements are less commonly used in research and industry because they are difficult to work with and extract. And for these reasons they are often far more expensive too. Many of the properties of the materials that they are made with can be made with cheaper alternatives. However, there are some specialist applications for the Ln elements. For example, the element neodymium is used in lasers and commercial magnets with very strong magnetic fields, and samarium is used for its magnetic properties.

Lanthanide phosphorus compounds are used in cathode ray tubes (CRT) that are found in television screens. They're used to give a variety of colors based on the compounds that are present and made with terbium and holmium. For example europium and yttrium compounds make red colors. They're used because of the unique fluorescence and luminescence properties; this is what gives them color.

You may find lanthanides as the flint in cigarette lighters; for example, alloys of cerium and iron form a material called auer metal. When it's struck, it makes a bright spark.

Feelin' Radioactive: The Actinides

The actinide elements begin with actinium. Actinium serves as a starting point for the other actinide series elements, which span 14 spaces across the periodic table from actinium to lawrencium. Actinides are more reactive than lanthanides. They form covalent compounds and organometallic complexes. Most of the actinide elements were discovered during development of the nuclear bomb. The set of actinides is represented by the symbol An.

The radioactivity of some of these elements causes higher reaction rates when binding other molecules. Otherwise, the actinides act almost exactly like regular transition metals (see Chapter 13). However, only a handful of examples exist because they're also typically radioactive and extremely difficult and dangerous to deal with.

Finding or making actinides

The half-lives of actinide elements span from tens of billions of years (since the birth of the universe 13.75 billion years ago) down to only a few hundred days. Thorium-232 has a half-life of 1.39×10^{10} years, and the shortest lived is curium-244 with a half-life of 162.5 days. (See Chapter 4 for details on half-lives and radioactive decay.)

Only four of the An elements are formed naturally: actinium-227, thorium-232, protactinium-231, uranium-235, and uranium-238. Notice that the element uranium is listed twice because they are two different isotopes (see Chapter 4 for a recap about isotopes). Some of the actinide elements were formed due to the highly energetic explosion at the start of the universe. For example, uranium-235, uranium-238, and thorium-232 have such long half-lives that the sources we find here on earth could have been created in the big bang.

The other more short-lived actinides must be made synthetically by using high-energy collisions in a particle accelerator. These machines collide a particle such as a gamma ray with an atom of the naturally formed actinide elements. They split after collision; the other elements are formed in the process of radioactive decay. The first of these new elements were named after the planets in a similar fashion to uranium (planet Uranus) — neptunium (Neptune) and plutonium (Pluto). The rest have been named for historical themes or places in which they were first created.

The quantities of these materials vary from several tons being recovered in a year, to others that have radioisotopes existing only a few micro seconds so there isn't enough time to collect large quantities of them.

Examining electronic structure

In the actinide series, the atoms add electrons to both d and f orbitals. Because the electrons join the d orbitals in ways similar to the 3d and 4d transition metals, they tend to undergo similar bonding arrangements as those 3d and 4d atoms. However, the arrangements found for organometallic compounds can have higher coordination numbers with actinides than with the transition metals. (See Chapter 9 to learn about the way that coordination complexes work.)

Actinide elements fill both 6d and 5f orbitals. The d orbital electrons bond much easier than the f orbital electrons, which is why the actinides are more reactive than the lanthanides (which fill f orbitals only).

The electronic structure of the actinide series starts with actinium [Rn] $6d^17s^2$ and continues up to lawrencium $[Rn]5f^{14}6d^17s^2$. This may appear complicated at first glance, but if you break it down you may find it easier to grasp the general changes in electron orbitals for atoms on the left side compared those of the right side. This represents the changes in electronic configuration going from actinium to lawrencium:

$$6d^1 \rightarrow 6d^1$$
$$0f^1 \rightarrow 5f^{14}$$
$$7s^2 \rightarrow 7s^2$$

Additional electrons can fall into either of the d, f, or s orbitals, depending on which has the lowest energy requirement. This causes some fluctuation in the chemical and physical properties of the actinide elements.

Actinium has a [Rn] $6d^17s^2$ configuration in the ground state, and for the most part, the 4f orbitals are typically more stable than the 5d orbitals. In the actinide series, it's common that with the addition of four or five more electrons, the 5f orbital becomes the more stable configuration.

Actinide elements typically have an oxidation state of either M^{3+} or M^{4+}. The reason we say "typically" is because not all of the elements have known oxidation states; they're dangerous and too short lived to be used for chemical reactions.

In these oxidation states, the ionic radii spans from $1.26 \angle$ (M^{3+}) to $0.96 \angle$ (M^{4+}). These are, on average, larger than the ionic radii found for the lanthanide elements, but because of the high capacity for bonding afforded by the use of d-orbitals, the bond lengths are much shorter, highlighting how the materials they make are more dense.

Atoms that have radioactive isotopes are very energetic because they're trying to reach a stable and steady state. (See Chapter 4 for more details on nuclear chemistry). For example, both actinium and curium glow in the dark, releasing light energy. For those An elements that are very radioactive, they can be used to form hydrogen gas. Many of the metals form hydrides quite readily, and the hydrogen atoms that crowd around the radioactive nucleus can be affected by the emission of alpha/beta/gamma rays. This can cause the hydrogen that's in H_2O to split from the oxygen and form hydrogen gas.

Comparing Reactivity: Actinide versus Lanthanide

The actinide elements undergo bonding very differently from the lanthanides due to the difference in the spatial extension of the frontier orbitals compared with the inner electron orbitals.

In actinides, the distance between the outer (valence electrons) orbitals and the inner electron orbitals is much greater when compared that of the lanthanide elements. In actinide, the distance between the 5f orbitals (outer orbital) and the 7s7p orbitals is greater than the distance between the lanthanide 4f orbitals and the 6s6p orbitals.

Because the 5f electrons have a much longer spatial extension than the 4f electrons of the lanthanides, the 5f electron orbitals can overlap with the orbitals of other atoms much easier, making them much more accessible for chemical reactions.

Looking More Closely at Uranium

Uranium gets a special up close and personal look because it's an important industrial material. It's used in nuclear reactors to make energy, and it also has uses for making nuclear weapons. Radioactive isotopes of uranium can be difficult to deal with when in their raw or metallic form; they are best handled when they are complexed or made into other compounds.

Binary compounds can be made with uranium. Such solids state compounds have been investigated because they have interesting magnetic properties. They are made by direct interaction with uranium metal. Oxides mainly form with the general formula UO_2, U_3O_8, UO_2. The metal also reacts with other elements such as boron, carbon, nitrogen, phosphorus, and arsenic to make semi-metallic solids. Compounds can also be made using silicon, sulfur, selenium, and tellurium. Urinates can be formed by the addition of uranium with alkali and alkaline Earth metals.

Halides are also formed by the addition of fluorine to make tetrafluorides (UF_4), pentafluoride (UF_5), and hexafluorides (UF_6).

Uranium can form a large amount of compounds with boron, particularly when in the oxidation states of either +3 or +4, such as $U(BH_4)_4$ and $U(BH_4)_3$.

Compounds of uranium can form with the pnictide elements, particularly nitrogen and phosphorus. There are many compounds known where nitrogen atoms are used as ligands. Oxygen and sulfur ligands can also be used for bonding with uranium; they can form a large range of alkyl and aryloxide compounds where the uranium has the oxidation state of U^{3+} through U^{6+}. One such oxygen containing ligand includes acetylacetonate (acac for short), that can form an eight coordinate complex of the form $U(acac)_4$. Thiol ligands have long been known for use with uranium, too.

Uranium hydrides (UH_3) is best used for chemical reactions because it decomposes slower than the uranium when it's in bulk metal form. The hydride decomposes at high temperatures to yield highly reactive uranium metal pieces.

Uranium has a strong capacity for bonding with water molecules, too. When they do this, the hydrolytic reactions lead to molecules that look like polymeric chains when carried out under the appropriate conditions. Aqueous solutions of uranium salts can easily undergo hydrolysis, and they have differing acid reactions that increase in activity from $U^{3+} < UO^{2+}_2 < U^{4+}$. A solution of U^{3+} in a 1 molar solution of hydrochloric acid can be stable for many days.

Uranium can also be used for mechanical properties, not just the radioactive properties. Depleted uranium is a form of uranium that has no radioactivity, or at least, has negligible radioactivity. The atom is very dense so any material made using this can be mechanically very strong. It has found favor for many military applications such as the skin of tanks. The depleted uranium can re-enforce the metal making it resistant to artillery fire. It can also be used as ballasts on ships, too.

Part IV
Special Topics

"So what if you have a Ph.D. in chemistry?
I used to have my own circus act."

In this part . . .

Inorganic chemistry is an extensive topic and doesn't stand alone. In the following chapters you learn how inorganic chemistry impacts other fields of chemistry and what kinds of things are possible when you apply the basics of inorganic chemistry to solve practical problems.

This part gives you all kinds of information, from how to break down poisonous gases so they don't pollute the air to how medicines affect illness without harming the rest of your body. The chapters in Part IV take you through the interesting special topics of organometallics, catalysis, bioorganic chemistry, and the study of solid state materials.

Chapter 15

Not Quite Organic, Not Quite Inorganic: Organometallics

Chemical reactions involving carbon are usually considered organic chemistry, while reactions using metals are usually defined as inorganic chemistry. In this chapter, you learn about the organometallics, which make some chemistry not quite organic and not quite inorganic, but rather a combination of both.

Metal complexes and compounds make interesting shapes that affect the material properties. Fortunately, we know a lot about them because of their use in catalysis (see Chapter 16 for details on catalysts and catalysis). In this chapter, we describe the bonding that occurs when carbon bonds with metals through covalent bonds.

Building Organometallic Complexes

Eighty percent of the periodic table is made up of metals. Generally speaking, metals are *electron deficient,* and the properties of metal compounds are a consequence of delocalized bonding of the electrons around the metal. In organometallic (OM) chemistry, metals that are electron deficient try to grab electrons from organic (carbon-containing) molecules so that they can become stable. The number of electrons a metal atom tries to grab and the number of electrons a ligand can share determine the shape and structure of the organometallic complex. This also affects the reactions that the organo-metallic complex can undergo. There's a sharing of electrons when the bond-ing is covalent, and there's a directionality to the ligand with respect to the metal center.

The organic compounds that bond with metals to form organometallic com-pounds are made of combinations of carbon, oxygen, nitrogen, and hydro-gen. Small molecules made from these atoms are called *ligands*. (Ligands are discussed in some detail in this chapter, but you can find more information about them in Chapter 9.) In essence, ligands bond around a metal center by donating electrons to the metal atom. Ligands come in various shapes and sizes; larger ligands called *cryptands* can also bond with metals, especially to the metals of Group 1 and Group 2. (These complexes are studied in the field of *supramolecular chemistry*.) See Chapter 9 to read more about the interest-ing complexes formed using Group 1 and Group 2 atoms.

The fact that bonding occurs via covalent interactions makes organometallics unique within the field of coordination complexes, specifically dealing with ligands that allow for bonding in a synergistic manner between the metal and the ligand. In particular that the Metal → Ligand occurs via σ-bond, or that the Ligand → Metal occurs via π electron flow. More details about covalent bonding can be found in Chapter 6.

Before you learn about how ligands bond to organometallic complexes, you need to understand the various electron rules that chemists use to predict the characteristics and behavior of metals when forming bonds.

Adhering to Electron Rules

Counting the number of donating or accepting electrons between metal atoms and ligand molecules can help you quickly learn to predict the stabili-ties, shapes, and structures of thousands of different organometallic com-pounds based on some simple calculations.

All the electron counting rules are used to determine how many electrons are required to make a stable compound. To achieve stability, the outer shell of an atom should have no vacant sites where electrons can go; in other words, it should have no free valence electrons, at which point it becomes stable and inert. When this occurs, the atom has the same electronic configuration as the next highest noble gas, and so becomes stable in a similar way to how the noble atoms are stable and inert.

All atoms in the periodic table, with the exception of the noble gasses, have some amount of reactivity associated with them. The reactivity occurs due to the sharing, losing, or gaining of electrons. In organometallic chemistry, the first place to start is with the metal. Consider it as a pure metal, and look at the valency. The *valency* describes the number of electrons that it has in the outer electron orbitals. These are at the frontier of the atom and determine how it reacts with the environment and with other atoms, or molecules such as ligands.

A great deal has been discovered about the exact structures of organometallics because the restrictions on the number of electrons around the metal center affords a high degree of control. For example, the precision to which chiral catalysis can be predicted and carried out is possible because no complex is possible, or stable, if it breaks the electron counting rules. (Chiral catalysis is discussed further in Chapter 16.)

Counting to eight: The octet rule

The octet rule works only for atoms with an atomic number less than 20 because at atomic number 21, we enter into the part of the periodic table that houses the transition metals. The transition metals (described in more detail in Chapter 13) are characterized by partially filled d-orbitals, and *d-orbitals* can hold up to ten electrons each. Each transition metal element has s-orbitals (space for two electrons), p-orbitals (with space for eight electrons), and d-orbitals (with space for ten electrons). Therefore, to make a transition metal stable, it must have a total of 18 electrons to match with the next noble gas configuration.

Transition metals are electron deficient and try to bind with molecules that have electrons to donate. They need to bind with as many donating groups as they can find to reach a total of 18 electrons in the complex.

Whereas the octet rule describes the basic philosophy behind the electron counting rules, the 18-electron rule is more useful when dealing with organometallic chemistry.

Calculating with the 18-electron rule

The *18-electron rule* is designed around the group number of the metal at the center of the compound. It bonds with as many ligands as needed to reach a total of 18 valance electrons. By doing so it achieves stability by reaching the electronic configuration of the next highest noble gas. For the most part organometallic compounds are restricted by the 18-electron rule, and this narrows down the possible ways that atoms can be bonded together.

Essentially, this rule allows you to predict what formulations create stable organometallic complexes. In short, when you know the number of valence electrons of the central metal atom (the group number), you can quickly back calculate what the required number of electrons are to fulfill the 18-electron rule. That is to say, how many more electrons it requires to reach a total of 18 electrons. If you know how many electrons are required, you can then find the appropriate ligand or combination of ligands that donates the required number of electrons.

The nucleus of an atom is surrounded by a series of an ever increasing number of orbitals around it. The first orbital holds two electrons and is then full and stable. The second and subsequent orbitals hold up to eight electrons each. When thinking about the stability of the organometallic compounds, think of the whole complex as a system that *acts* like a nucleus when trying to achieve stability.

There are two conventions used when counting electrons: ionic and covalent (sometimes called the *radical method*). Consider the organic molecule cyclo-pentadiene, a planar five-sided hydrocarbon with the chemical formula C_5H_6. This is the precursor for the cyclopentadienyl ligand, also a planar molecule but has the chemical formula C_5H_5 and is given the abbreviation Cp.

The cyclopentadienyl ligand can be considered as ionic in the case where $Na^+C_5H_5^-$ is formed. This occurs by the removal of a proton from C_5H_6. But Cp can also be considered as a neutral radical $C_5H_5\cdot$. This is known as the covalent (radical) convention. A radical is a species containing at least one unpaired electron, and the dot is used as a reminder that it's a radical.

It doesn't matter which convention is adopted, as seen in Table 15-1. Both conventions count the same total number of electrons; for example, 18 electrons. The choice does affect the assigned formal charge on the central metal atom— d^6 is for Fe^{2+}, and d^8 is for Fe^0. In practice either convention can be employed (d^6 represents that there are six electrons in the d-orbitals).

Table 15-1	Electron Counting as Either Ionic or Covalent		
Ionic Convention	*Electron Count*	*Covalent Convention*	*Electron Count*
$C_5H_5^-$	$6\ e^-$	$C_5H_5\cdot$	$5\ e^-$
$C_5H_5^-$	$6\ e^-$	$C_5H_5\cdot$	$5\ e^-$
Fe^{2+}	d^6	Fe	d^8
Total electrons	$18\ e^-$	**Total electrons**	$18\ e^-$

As with any rule, there are exceptions. For example, cluster compounds don't always satisfy the 18-electron rule. And some organometallic compounds find stability with only 16 electrons instead (more on this in the next section). But the 18-electron rule is a good place to start.

Settling for 16 electrons

The *16-electron rule* is an extension on the same simple idea expressed by the 18-electron rule — that is, that for an organometallic complex, stability is reached when the valence electrons sum to a certain value. In this case, that value is 16 electrons.

For one of several possible reasons a metal center may accept only 16 electrons total, so another ligand with two more electrons can't fit. This can occur when the central atom is too small to fit enough electron-bearing atoms or molecules around it; it may find relative stability with only 16 electrons, or it may be a transition state in a catalytic reaction.

Some more stable examples include IrCl(CO), commonly known as *Vaska's complex*. Sometimes the atoms that are the source of the electrons are too large to fit around the smaller metal atoms, and so a 16-electron rule is more applicable. Or in rare cases such as $V(CO)_6$ the metal center is so small it can't fit a seventh coordination site.

Exceptions to the 18-electron occur, for example where 16 electrons is a meta stable transition state in catalytic reactions, but these are often very short lived intermediate species. In a catalytic cycle the catalyst toggles back and forth between an 18 electron and 16 electron species.

Effectively using the EAN rule

In addition to the 18- and 16-electron rules, there is also the Effective Atomic Number (EAN) rule, which looks to the periodic table for verification of what can be calculated using the 18-electron rule. To put this rule simply, it says that an organometallic complex (as a system) binds as many ligands as possible to reach the number of electrons of the next noble gas configuration. For example, for carbon the noble gas configuration is that of neon and it's achieved with a total of ten electrons [Ne]. For iron, the next noble gas configuration is krypton, and it has 36 atoms [Kr].

Consider the example of iron pentacarbonyl ($Fe(CO)_5$) that's used in the catalytic growth of nanotubes. Table 15-2 illustrates how both the 18-electron and the EAN rules can predict a stable configuration. In this example, the covalent convention is used.

Table 15-2   Comparing the EAN Rule with the 18-Electron Rule

Effective Atomic Number Rule	Electron Count	Covalent 18-Electron Rule	Electron Count
Fe	26 e$^-$	Fe	8 e$^-$
5*CO	10 e$^-$	5*CO	10 e$^-$
Total electrons	36 e$^-$ [Kr]	**Total electrons**	18 e$^-$

When using the 18-electron rule, you should count the number of metal electrons according to the group number. For example, iron is in Group 8 of the periodic table, so you give it eight electrons. But for the EAN, the number 26 is used instead because this is the atomic number of iron.

Bonding with Metals: Ligands

Organometallic compounds are systems that have ligands attached to a metal center. The ligands form bonds with the metals and can also help connect them to other metals by acting as a bridge between two metal atoms. The word *ligand* originates from Latin, meaning "to bind" and refers to the fact that these organic groups bind to metal centers.

They do this based on Lewis acid-base chemistry, which is another way of saying it's about accepting and donating electrons. (See Chapter 5 for a refresher on acid-based chemistry.) The ligands act as a Lewis base because they're an electron pair donor. The central metal atom acts as a Lewis acid because it's an electron pair accepter.

The number of ligands bound to the metal ion is designated as the coordination number of the metal. Some main group elements form complexes that are similar to transition metal complexes. These include aluminum, tin, and lead, for example, but for the most part organometallic chemistry employs transition metals. This is due to the partially-filled d-orbitals.

When one bond is formed between the metal and the ligand, it's known as *monodentate;* with two bonds it's known as *multidentate*. Multidentate compounds form a class of materials known as *chelating complexes*. (For more details about this see, Chapter 9 about coordination complexes.) Think of a chelating molecule like a finger and thumb that can hold onto a tennis ball.

If the number of ligands present is not sufficient to fulfill the electron counting rules, then cluster compounds can form. This also occurs if metals are not able to satisfy the 16-electron or 18-electron rule. This can force metals to bond with each other, so you have metals bonded to one another (often with a strong double bond) that are stabilized by a series of ligands around them.

There are many types of ligands, but the most common are carbon based and are called *carbonyls*. The carbonyls provide an excellent example of how ligands bind to react with organometallic complexes.

Including Carbon: Carbonyls

A *carbonyl* is a functional group that is composed of a carbon atom double bonded to an oxygen atom (CO). Carbon monoxide can bind either through the oxygen or the carbon atoms, and as such it can also act as bridge between two metal centers. One metal center can be bonded to the C, whereas the other be bonded to the O.

For example, the oxygen in CO acts as a donor to Lewis acid (electron acceptor) compounds such as aluminum trichloride, $AlCl_3$ and in doing so bridges the two metal centers (M – CO – M). This causes a weakening of the C – O bond because it's stretched, which may result in the breaking of the C-O bond, too. This capacity for stretching bonds and then breaking them is what makes transition metals so important in catalysis.

There are multiple types of carbonyl, described according to which elements they include, or how many ligands are attached to the metal.

- ✔ **Binary carbonyls:** These compounds have a metal center and a carbonyl ligand. They are the simplest carbonyl compound.

- ✔ **Isocarbonyls:** These compounds have one carbonyl ligand that is bonded to two metal centers at the same time, because it has both carbon and an oxygen available for bonding to the metal (*iso* means two).

- ✔ **Binary metal carbonyls:** These compounds, also called *homoleptic carbonyls*, have the general formula: M(CO)n.

A homoleptic complex is one in which all the ligands are identical to each other. In the case of homoleptic carbonyls, all the ligands are carbonyls. This can be compared with heteroleptic complexes that have more than one ligand species in the organometallic compound — one ligand is CO and the other can be a CN for example.

There are also many mixed carbonyl compounds, but the binary compounds were known first. Of these, the first known was $Ni(CO)_4$, discovered by Ludwig Mond in 1890, and from this the reaction by which nickel can be purified is referred to as the Mond process. A fascinating aspect of carbonyl chemistry is that none of the main-group elements form metal carbonyls, so there are no zinc or magnesium derivatives.

Table 15-3 shows the binary transition metal carbonyls that you can find. As you can see there are more carbonyl complexes for the elements in the middle of the transition metal group, and that to the far left and far right there are far fewer possibilities (if any at all). This is because the central transition metals have a larger variety of oxidation states, with iron and cobalt forming the most amount of carbonyls.

Table 15-3		Binary Metal Carbonyls with Transition Elements					
Sc, Ti	*V*	*CR*	*Mn*	*Fe*	*Co*	*Ni*	*Cu, Zn*
	$V(CO)_6$	$Cr(CO)_6$	$Mn_2(CO)_{10}$	$Fe(CO)_5$	$Co_2(CO)_8$	$Ni(CO)_4$	
				$Fe_2(CO)_9$	$Co_4(CO)_{12}$		
				$Fe_3(CO)_{12}$	$Co_6(CO)_{16}$		

Providing the Best Examples

For the most part, organometallic ligands are made of either carbon or nitrogen. These two elements serve as excellent guides for what you can expect when using ligands. In this section, we describe how the atoms function as ligands, and we also describe boron, which behaves in a unique manner.

e-precise carbon

To reach a stable state, carbon needs a total of eight electrons (remember the octet rule!). The valence of carbon is four, so it's looking to gain four more electrons by sharing with some other atoms. This can be achieved by bonding to other carbon atoms, or by bonding to a multitude of other atoms from the periodic table. As a result, carbon compounds are the most numerous compounds known to scientists.

When carbon bonds with atoms from its own group (such as silicon, germanium, and so on), the molecules are known as *congeners*. This is a term given to atoms that are from the same group as carbon and have similar chemical structure and share similar properties. These similarities result in bonds that aren't very polar. Carbon ligands are typically composed of hydrogen, oxygen, or nitrogen bonded to carbon.

e-rich nitrogen

Nitrogen compounds such as cyanide (CN^-) or ammonia (NH_3) act as good ligands because nitrogen is a good donor atom and donates electrons readily to the metal center. The nitrogen group act as good Lewis-base compounds, and the Lewis basicity gets much larger with these compounds.

Many organometallic ligands are composed of *amides,* which are nitrogen-containing ligand molecules that are often bonded with carbon — and sometimes with oxygen.

When isocyanide ligands are attached to a metal center of strong π donation, they are great electron acceptors, similar to CO. Isocyanides are good Lewis bases; they can form bonds to M ions involving only σ-donation. CN^- is isoelectronic with CO and RCN, but it's not as good a π-acid as the others. For a recap on π and σ bonds see Chapter 6.

Ligands involving dinitrogen are far more inert and can bind only to a metal in the presence of other ligands. They can't make homoleptic complexes the way CO or RNC can.

The other elements in this group can act as π-acceptors, too, if the metal center is electropositive.

e- deficient boron

Interestingly, whereas boron is in the same group as carbon and nitrogen, it behaves differently when forming organometallic compounds. Carbon and nitrogen are strictly nonmetals, so they're treated as ligands strictly. But boron is a *metalloid.* This means that when forming compounds with carbon, boron may act as the metal or the ligand. For this reason, boron can form a number of metallic borides, organoboron compounds, and boron hydrides.

With three valence electrons, boron tries to find five more electrons to reach eight total electrons. With only three electrons, it's *electron deficient* and attempts to make as many bonds as possible to reach noble gas configuration. To do this, it can react with itself in ways that are unique and special. In fact, many of the structures that boron form can't be replicated using other elements.

To gain five electrons, boron builds unique *closo* or cage-like molecules. One class of neutral B-H ionic species called *boranes* serves as a good example of what could be possible with other organometallic compounds that form cage-like molecules.

These unique compounds have the formula $B_nH_n^{2-}$, where n represents the number of corners of the cage molecule. In each corner, the BH groups sit in the central plane of a triangulated polyhedron, which is illustrated in Figure 15-1.

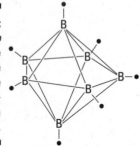

Figure 15-1:
Closo
borane
cage-like
structure
with chemi-
cal formula
closo-$B_9H_9^{2-}$.

Each boron atom is bonded to four other boron atoms and a hydrogen atom, which makes it stable because it has a total of eight electrons (follows the octet rule).

It can also bond to other ligands that have same number of electrons by bonding to CH groups, for example, and forming nido complexes. *Nido* complexes are described by the formula $B_9H_9^{2-}$ where one of the edges of the cage structure is missing, and some of the boranes are replaced by a CH group. This is possible because the borane (BH:) and CH groups each have the same number of electrons; they are *isoelectronic*.

Compounds where CH is inserted in place of BH are known as *carboranes* (or *carbaranes*). In practice, it's common to find these molecules with two carbon atoms instead of just one.

The diamondoid shape that the boron compound forms demonstrates how it can be bonded into a diamond crystal without causing strain in the dia-mond lattice due to its similarity of structure. This also demonstrate how the properties of a material at the macro scale (visible to the naked eye) is deter-mined by bonding at the atomic scale.

Behaving Oddly: Organometallics of Groups 1, 2, and 12

Groups 1, 2, and 12 can be thought of as two parts — Groups 1 and 2 elements comprise the first part, and the late transition metal atoms for Group 12 comprise in the second part. The reason to break it down like this is simple. On the one hand, Group 1 and Group 2 atoms are basically main group elements that can form several organometallic complexes. On the other hand, it should be evident by now that organometallic complexes are predominantly formed by transition metals, but these transition metals in Group 12 are a little different. They don't tend to form organometallic complexes because they don't have partially filled d-orbitals. Thus, the reactivity of these atoms is very different, to the extent that they are quite inert. This is why they're often used as coinage materials.

All alkali metals have organometallic derivatives. One example is alkyl lithium, which can be used as a catalyst to form carbon-carbon bonds. This is done by the following the reaction, where R is some otherwise unreactive material.

R-X + R-Li = R-R

R is used in chemistry to denote a molecular group. It can be the same group, or different groups R_1 and R_2.

This is useful because it catalyzes the reaction between carbon atoms. Grignard reagents using organomagnesium are often used to aid in organic chemistry synthesis in the reaction to bond C-C bonds.

Chemistry of the of the s-block elements is usually carried out in nonaqueous environments, because they react so vigorously with water. The organic ligands stabilize them and keep them from reacting explosively.

The Group 12 atoms are on the edge of the transition metal part of the periodic table, and because these elements don't form any compounds in which the d-orbital shell is not full, they're not considered a transition metal. They do, however, resemble the d-group elements in their ability to form complexes with amines, halide ions, ammonia, and even cyanide. In the case of bonding with cyanide, CN^-, the possibility for $d\pi$ bonding between the metals and the ligand is much lower when compared with the rest of the transition metals.

Developing modern silicone

Alkyl groups can be bonded to Group 14 elements such as silicon (Si) using Grignard reagent or alkyllithium reagents. This produces chloromethylsilanes that are commonly used to make oily products and rubbers. But the use of something like alkyllithium isn't convenient on an industrial scale. In 1940 a method was developed using pure Si reacting with an alkyl or aryl halide over a copper catalyst. This reaction led to the widespread access of silicon polymers, such as silicones, for example. Prior to this, the most facile route was to employ naturally formed rubber that would come from the sap of rubber trees, for example this process of forming natural rubber is still being carried out in countries like Thailand. During World War II, the requirement for rubber increased dramatically, and the capacity to create silicones by this route made the large scale supply of silicones more attainable. The formation of silicone is carried out according to the following equation:

$$Si + RX \rightarrow R_n SiX_{4-n}$$

Sandwiched Together: Metallocenes

Metallocenes are organometallic coordination compounds composed of a metal center that's bonded to the face of planar closed loop hydrocarbons, such as cyclopentadienyl ligands. These are more commonly known as *sandwich molecules* because the metal is sandwiched between the two planar molecules. They have the general formula $(C_5H_5)_2M$.

Metallocenes are important forms of organometallic chemistry because they have remarkable stability, which at first was not understood. This led to further study, making metallocenes important historically in developing the theory and practice of modern organometallic chemistry.

The most common of the metallocenes is ferrocene. The nonbonding and bonding orbitals are exactly filled, and so they form the most stable compounds, followed by compounds made of other Group 8 elements. But Group 9 and 10 have one or two electrons in anti-bonding orbitals, so cobaltacene, for example, is an 18 electron cation, $(C_5H_5)_2Co^+$.

Metallocenes with extra electrons in antibonding orbitals are also paramagnetic. Paramagnetic metallocenes are very reactive; the most reactive is $(C_5H_5)_2Ni$. (See Chapter 9 to read more about magnetic properties.)

It's possible to bond three cyclopentanes with elements of Group 4, and with the heavier elements of Groups 5 through 7. This causes the cyclopentanes to be strained away from each other, causing the planar molecule to curve a little (this effect is known as *steric strain*).

The most important use of metallocenes is for polymerization of alkanes. In particular, one can make a form of polyethylene that is linear and has a low density. The low density is achieved because there are no branches formed, which is not common in catalysis because other catalysts make branched polymers more often.

Clustering Together: Metal-Metal Bonding

When metal elements bond to one another, they form metal-metal bonds with unique molecular structures. Metal-metal (M-M) bonds are grouped into four types: edge-sharing bioctahedra, face-sharing bioctahedra, tetragonal prismatic structures, and trigonal antiprismatic structure. Compounds with edge-sharing are the most numerous and include M-M bonds with up to three other metal centers.

When metals come together and bond in conjunction with bonding of ligands, they look like a cluster of metal atoms in the center with the ligand dangling around them. For this reason, they're known as *cluster compounds.*

There are two main classes of cluster compounds — those where the metal atoms have a low formal oxidation state, in which case the ligands are almost always CO groups; and those with metals that have a high oxidation state, ranging from +2 to +4. In this case, the ligands are usually halides, sulfides, oxide ions, or ligands from a monomolecular Werner complex (see Chapter 9 for details on a Werner complex).

Cluster compounds were first discovered in 1935 using X-ray crystallography of a tungsten compound (potassium nonachloroditungstate) $K_3W_2C_{19}$, as shown in Figure 15-2. In normal metal bonding, the W-W bond length is 275 pm, but in this crystal it was found to be 240 pm. Shorter bond lengths lead to stronger bonds.

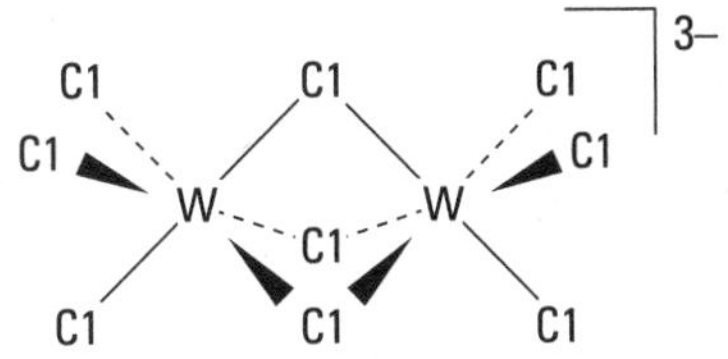

Figure 15-2:
Tungsten
cluster
compound.

Tetramethyldiarsanse $(CH_3)_2AsAs(CH_3)_2$ is one of first organometallic compounds to be made. It's a catenated compound based on As-As bonding.

Metal-metal bonding in organometallic compounds can induce double bonds to form between the metal atoms. This isn't common for metallic bonding because normally metallic bonding is known for having an excess of delocalized electrons that are not associated with any bonding. (Delocalized electrons are what make metals metallic; see Chapter 8 for more details.)

Creating Vacancies: Insertion and Elimination

There are many reactions that organometallic compounds can undergo, but the most important from an industrial perspective are those of insertion and elimination. These kinds of reactions are how many materials are synthesized and is the mechanism by which many catalysts work also.

Take the example of a CO insertion to the manganese complex $Mn(CO)_5CH_3$, which is illustrated in Figure 15-3.

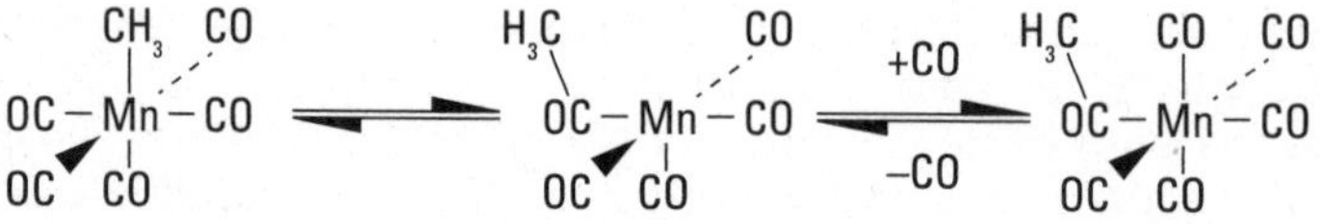

Figure 15-3: Insertion of carbonyl group around a crowded metal center is achieved by the migration of an alkyl group that creates a vacancy for carbonyl insertion.

In this reaction a CO becomes inserted between the Mn metal center and the rest of the ligand, as follows

$$Mn(CO)_5CH_3 + CO \rightarrow Mn(CO)_5C(O)CH_3$$

In this case, the word "insertion" can be a little misleading because the carbonyl group that instigates the reaction is not the CO that's inserted into the compound. It is a two-step process that has a 16 e- transition state. In the first step a vacancy opens up when the methyl group (CH_3) moves to the side and bonds to the terminal CO group instead of to the metal center. This is known as an *alkyl migration*, and the transition state is a 16-electron compound. In the second step the vacancy is then filled by another CO ligand, thus creating a Grignard reagent.

Other insertions can occur like in the case where unsaturated ligands such as an alkane can insert between M-C or M-H bonds, this is highlighted in the following example:

$$L_nM\text{-}H + CH_2\text{=}CH_2 \rightarrow L_nM\text{-}CH_2\text{-}CH_3$$

Remember that many reactions are reversible and that the backward reaction of insertion is *elimination*. The reverse reaction of this insertion, when the alkane is removed, is called the β-*hydride elimination reaction*.

Many of the reaction mechanisms are important to catalysis and many organometallic compounds are used widely as homogenous catalysts. Some of the reactions include catalytic alkane polymerization, oxidative additions, and reductive elimination. Although these are very specific reactions with different mechanisms used to describe each, you should be able to understand them by ensuring the electron counting rules are obeyed.

Synthesizing Organometallics

There are a number ways in which organometallic compounds can be made. In general, they fall into one of two categories:

- ✔ **Reactions involving the gain or loss of a ligand:** These reactions deal with the addition or subtraction of a ligand to or from the metal center and include ligand dissociation and substitution, oxidative addition, reductive elimination, and nucleophilic displacement.

- ✔ **Reactions involving the modification of ligands:** These reactions modify the present ligands, without adding or removing any. These include insertion reactions, carbonyl insertions (also known as alkyl migration), hydride elimination, and abstraction.

A list of some common organic ligands that are used in organometallic chemistry is given in Table 15-4, which shows several ligands named along with their chemical formula. This table also shows the number of electrons that are donated by the ligand. The hapticity value tells you how many bonds there are between the metal and that ligand. Some ligands are bonded in only one location with the metal, but several ligands can bond with the same metal atom through multiple bonds. For example, if a ligand is bonded to a metal in two places then this is denoted in chemical equations as η^2.

Table 15-4   Common Ligands Used in Organometallic Complexes

Ligand Name	Ligand Formula	Hapticity	Electron Number
Methyl, alkyl	CH_3, RCH_2	η^1	1
Alkylidene	R_2C	η^1	2
Alkylidyne	RC	η^1	3
Ethylene (ethane)	C_2H_4	η^2	2
Allyl (propenyl)	CH_2CHCH_2	η^1	1
		η^3	3
Cyclopentadienyl	C_5H_5	η^1	1
		η^3	3
		η^5	5
Benzene	C_6H_6	η^6	6

One method for creating an organometallic compound uses a metal salt as a precursor. The metal salt is used as a starting material ($CrCl_3$) and is reacted with a ligand of choice, in this case cycolopentane. Aluminum metal forms a much more stable chloride than the chromium, and this releases the chromium, making it available for bonding with the cyclopentane. In the metal salt reduction, the presence of the ligand can cause metal to bond to ligand such as in the following equation.

$$CrCl_3 + Al + 2\,C_6H_6 \rightarrow AlCl_3 + [Cr(\eta^2\text{-}C_6H_6)_2]$$

This can occur for reactions of transition metal salts with a main-group organometallic compound. For this type of reaction it's common to use the sodium salt $Na^+(C_5H_5)^-$ to deliver C_5H_5.

$$FeCl_2 + 2\,Na^+(C_5H_5)^- \rightarrow 2\,NaCl + [Fe(\eta^2\text{-}C_5H_5)_2$$

This can also be achieved using a Grignard reagent or an aluminum alkyl such that:

$$WCl_6 + 3\,Al_2(CH_3)_6 \rightarrow W(CH_3)_6 + 6\,AlCl(CH_3)_2$$

Another way in which an organometallic compound is made is by reacting a raw metal in the vapor phase with free ligands in the vapor phase also. To produce the vapor phase, the metal is heated to such an extent that it sublimates, and turns from a solid to a gas. This can be done in a furnace where the vaporized metal passes to a condensation stage where upon the ligand condenses on the metal center. Gas phase reactions can be done at very high temperatures, and so it's possible to make organometallic compounds that can't be made in the liquid phase.

$$Ti\,(g) + 2\,C_6H_6 \rightarrow [Ti(\eta^6\text{-}C_6H_6)_2]$$

Showing Similarities with Main Group Chemistry

Both organometallic compounds, with metal centers and multiple ligands bonded around them, and *elemental hydrides* have a distinct similarity. They are said to be *isolobal* to each other.

Elemental hydrides are molecules formed by a main group element bonded to multiple hydrogen atoms, such as methane CH_4.

The isolobal analysis looks at the ligand and the hydride, and compares the outer electron shells, or frontier orbitals, of each. The *frontier orbitals* are usually the ones participating in the chemistry. If the frontier orbitals have the same electron count and the same orbital type and symmetry, they're most likely isolobal to one another and will react similarly.

For this discovery Ronald Hoffman and Kenichi Fukui were awarded the Nobel Prize in 1981. They both independently discovered that the symmetry of the ligands in both cases of ligands of elemental hydrides and organometallic complexes could be used to understand chemical reactions. Specifically they suggested that the frontier orbitals could be used to predict certain reactions that had previously been difficult to understand. They suggested that for organometallic compounds with a hydrocarbon anion as a ligand, that the ligand is similar to, or isolobal to, the hydride anion.

To be precise, if the number of orbitals, the symmetry of the orbitals, the energies, the shapes, and the electron occupancy of these orbitals are similar, the ligand and the hydride are considered isolobal. The isolobal relationship is denoted using an upside down lobe that is placed below the double headed arrow, as can be seen in Figure 15-4.

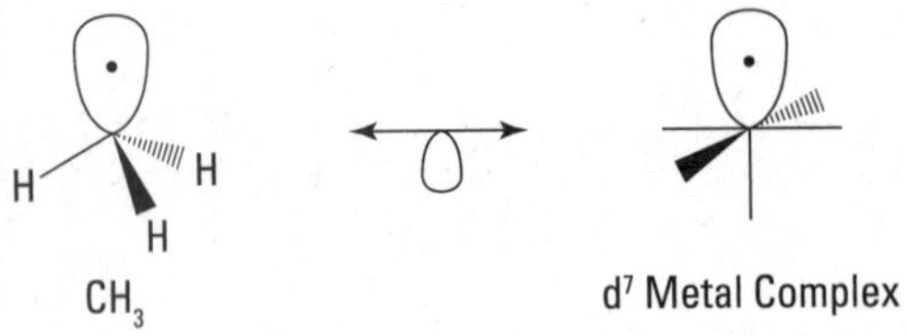

Figure 15-4:
Isolobal
principle
between a
CH_3 elemen-
tal hydride
and a d^7
metal
complex.

The isolobal theory has practical application because it suggests that isolobal fragments of compounds (the ligand and hydride) should have similar reactivities, and in principle can be exchanged in a molecule without changing the bonding arrangement of the molecule. An example of how this can be applied is with methyl and ethyl derivatives. Whereas they may have different properties, if you need to do a reaction in an aqueous solvent instead of an organic solvent, you might want to change the methyl groups out and put in hydrides instead. And you can do this expecting that the overall stability properties of the organometallic compound won't change significantly. This should work in principle, but slight subtleties may occur, so in practice you have to test whether a substitution is possible, and then see what effect it makes to the compound.

Chapter 16

Accelerating Change: Catalysts

. . . .

In This Chapter

▶ Speeding up chemical reactions with catalysis

▶ Producing materials from plastics to medicines faster and cheaper

▶ Using transition metal complexes as catalysts

. . . .

The old saying "Time is money" has never been as true as in the world of catalysts. Simply put, catalysts make materials affordable on the industrial scale. For example, just before WWI, catalysts allowed for the synthesis of ammonia gas from the reaction of hydrogen and nitrogen gases. This may not look very impressive at first glance, but remember that the two nitrogen atoms of nitrogen gas are bound together by a strong triple bond. In fact, despite the help of a catalyst, the world still uses about 1 percent of man-made energy creating the stuff.

What's so special about the pungent, household cleaner? The majority of ammonia produced is utilized to make nitrogen-rich fertilizers, but ammonia can also be used to make saltpeter ($NaNO_3$), a common precursor for both gunpowder and certain explosives. As you can imagine, during times of war, it would be useful to have the ability to grow more food with nitrogen-rich fertilizer *and* make more gunpowder. This is only one example of catalysts in history. The science of catalysis has boomed over the years, and much of what humans use today was developed as a result.

Speeding Things Up – The Job of a Catalyst

A catalyst helps change the speed of a chemical reaction. A catalyst can be either a positive catalyst that speeds up a reaction or a negative catalyst, commonly referred to as an inhibitor, which slows down a reaction. Positive catalysts, referred to simply as *catalysts*, are much more commonly sought after because they yield products much more quickly while requiring less energy, two factors that ultimately can lead to higher profits.

Figure 16-1 shows how a catalyst can provide an alternative pathway for the reaction to occur, it does this by allowing a transition state that requires less activation energy for the reactants to react and form products. Notice that the y-axis on the graph is a plot of the Gibbs energy. For both the reaction with and without catalyst present, the energy of the reactants and products are unchanged; however, the energy barrier required to overcome for the reaction to occur has been lowered for the catalyzed reaction. In turn, this can shorten the timeframe for the reaction to occur because now there is a higher likelihood that the reactants will collide with enough energy to overcome the energy barrier.

Figure 16-1:
A potential energy diagram showing how a catalyst works by giving alternative energy pathways in a reaction.

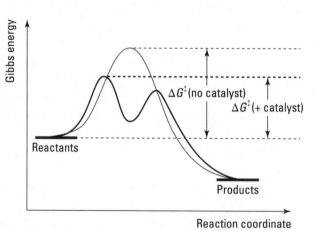

Although there are a few different reaction schemes, most catalysts work in a cycle where they regenerate themselves. For example, assume you have a relatively slow reaction that forms a product (Z) from two reactants (X and Y):

$$X + Y \rightarrow Z$$

If you add catalyst *(C)*, a possible reaction scheme could look like:

$$X + C \rightarrow XC$$
$$Y + XC \rightarrow XYC$$
$$XYC \rightarrow CZ$$
$$CZ \rightarrow C + Z$$

Notice, however, if you sum up these four reactions, putting all the reactants together and then all the products together, you get:

$$X + C + Y + XC + XYC + CZ \rightarrow XC + XYC + CZ + Z$$

Cancel out anything you see on both sides, and you are back to:

$$X + Y \rightarrow Z$$

A catalyst performs by altering the kinetics, or speed, of the reaction. When a catalyst turns one molecule into another, the corresponding speed is called the *turnover number* (TON). This measures how many molecules are being converted by the catalyst. In catalysis reactions where the binding and reaction sites are well understood, you can also determine the *turnover frequency* (TOF), which describes the speed, or Rate (n), at which the catalyst can work.

The Rate for a particular reaction is described by the following equations:

$$A \xrightarrow{\;c\;} B \qquad\qquad v = \frac{d[B]}{dt}$$

Here, reactant A is being converted to product B with the help of catalyst *C*. The rate of this reaction, v, is therefore the change in the concentration of product (B) over time. The TON (N) is defined by:

$$N = \frac{v}{[C]}$$

A highly active catalyst, or a catalyst with a large TOF, can cause a fast reaction even in low concentrations. It's significantly more difficult to determine the TOF of heterogeneous catalysts because the exact details of the reaction sites are still unknown. For heterogeneous catalysts with a solid catalyst, it is therefore easier to consider the surface area instead of *[C]*.

Catalysts are chosen based on their activity, selectivity, and lifetime:

✔ *Activity* measures the number of complete cycles the catalyst performs in a set period of time.

✔ *Selectivity* is based on exact products that are created. Examples where selectivity is important include molecules with a very discreet chain length and molecules with specific stereochemistry (3D structure).

✔ *Lifetime* expresses how long the catalyst lasts before it needs to be replenished or replaced.

A catalyst undergoes slight changes to move protons and electrons around, but after the desired product is formed, the catalyst is done with that molecule and begins to start the process anew with fresh reactants. This happens in a cyclical manner, leaving the catalyst continually ready to start over again. Unfortunately, whereas this is an impressive feat, catalysts don't make magic happen; they cannot make impossible reactions occur. Instead, for any reversible reaction, catalysts alter the rate of the reaction to achieve equilibrium. In essence, a catalyst speeds up both the forward and the reverse reaction equally. The final position of the equilibrium, however, stays untouched. So remember, reactions that are thermodynamically unfavorable cannot be made favorable by a catalyst.

Considering Types of Catalysts

There are two broad distinctions in catalysis that determine certain characteristics of the reaction and the products that are created:

- Homogenous catalysts are soluble in the reaction medium. These catalysts work under mild temperatures and pressures, yet have to be separated from the product later. Examples include enzymes that catalyze biological reactions and the use of acids and bases to catalyze the hydrolysis (breaking of a chemical bond by adding water) of esters.

- Heterogeneous catalysts are in a different physical state than the reaction medium, so they are easily separated from the product; however, they often work at high temperatures and pressures that are more energy intensive on large industrial scales. An example of a heterogeneous catalyst is a catalytic converter (solid) that catalyzes the transformation of toxic combustion products to less toxic carbon dioxide (CO_2), nitrogen (N_2), and water (H_2O).

In any catalytic reaction, the catalyst undergoes a cycle that changes the state of the catalyst, effectively to make room for the reactants, release the resulting products, and start all over again as shown in Figure 16-2. The steps that a catalyst can undergo include ligand substitution, oxidative addition, reductive elimination, and insertion reactions.

Homogenous catalysts

Homogenous catalysts are often composed of a transition metal with various ligands bonded around the metal center and are used because of their versatility. These catalysts can undergo many bonding configurations due to the wide variety of coordination numbers available to transition metals, giving rise to lots of bonding geometries. In practice, this means it can take reactants and move them around the metal center until they are in a good position for bonding to occur.

A needle threader is a tool used by people who sew; it helps them put a piece of thread through the eye of a needle. In short, it's a small piece of stainless steel wire that's rigid and easier to push through a needle's eye than a flimsy piece of thread. When the wire is through the eye of the needle, it forms a loop about the size of a dime (significantly larger than the eye of the needle). Now if you only have to get the thread through a hole the size of a dime as opposed to a needle's eye, you can thread a lot of needles much faster; however, at the end of the day, you start with a piece of thread and a needle and you finish with a threaded needle with or without a needle threader. The needle threader doesn't really do anything but simply change the geometry of threading a needle; this is similar to a homogenous catalyst that helps orient the reactants so that the reaction is carried out much faster and then it can return to its original geometry after the reaction is complete.

Homogenous catalysts often undergo a series of linked chemical reactions with each new stage forming a different metal complex. To do this, the reaction sequence goes through a variety of reaction equilibria because each intermediate produced is only pseudo-stable. The instability of these intermediates makes them reactive; therefore, an increase in the rate of the reaction occurs.

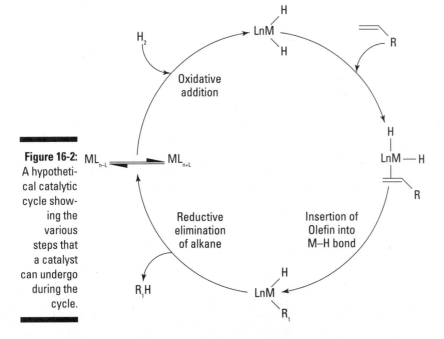

Figure 16-2: A hypothetical catalytic cycle showing the various steps that a catalyst can undergo during the cycle.

When using a homogenous catalyst containing a transition metal, one must consider the size of ligand that binds to the transition metal. The larger the ligand, the more it limits the space available for the bonding of reactants. Imagine a reactant flying into to a transition metal atom — the more room available to hit the atom, the more likely the reaction will occur. You can increase or decrease the room available (known as steric effects) by carefully selecting what ligands bind to your transition metal.

Ligands coordinated to metal centers undergo dramatic changes in reactivity, including CO, $R_2C=CR_2$, PR_3, and H^-. These ligands act as Lewis bases, and in turn, the metal ion acts as a Lewis acid. (For more on Lewis acids and bases, see Chapter 5). These ligands are effectively broken into two sub-groups: ionic ligands, such as Cl^-, H^-, OH^-, CN^-, alkyl$^-$, aryl$^-$, $COCH_3^-$, and neutral ligands, such as CO, alkene, phosphine, arsine, phosphite, H_2O_2, and amine.

Alkenes can be converted into very useful aldehydes by using a cobalt catalyst. A carbonylation reaction involves the addition of H_2 and CO to an alkene, as shown in Figure 16-3. Without the use of catalyst, this reaction runs far too slowly to be significant.

Figure 16-3:
Carbonyl-
ation
reaction.

$$\text{C}=\text{C} + \text{CO} + \text{H}_2 \xrightarrow{\text{Co}} \text{H}-\text{C}-\text{C}-\text{C}\overset{O}{\underset{H}{\diagdown}}$$

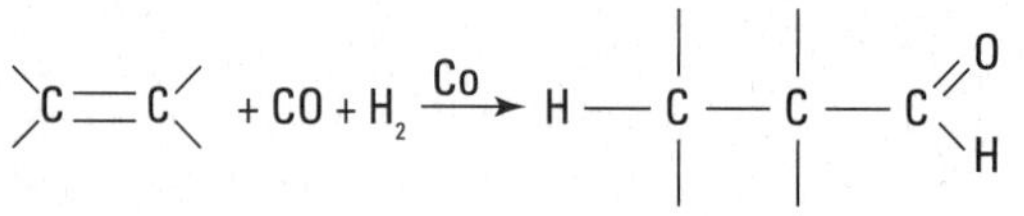

The following sections include some specialized uses for catalysts in an industrial setting.

Producing aldehyde groups (CHO) from alkenes is known as hydroformulation — an aldehyde and a hydrogen atom are added to a carbon-carbon double bond. Usually, the aldehydes aren't the final desired product; instead, the aldehydes are commonly treated with hydrogen to produce primary alcohols (RCH_2OH). Historically, hydroformylation is a key reaction to the development of the modern field of industrial chemistry.

Group 8 (Fe, Ru, Os) metals bound to carbon monoxide (CO), also known as metal-carbonyl complexes, can be used for carbonylation reactions, or reactions that use CO, involving alkynes, alkenes, or alcohols as substrates. Products include carboxylic acids and various derivatives, the process is generally known as *Reppe reactions*. Named after Walter Reppe, a German chemist noted for his contributions to chemistry through studies with the molecule of acetylene (properly called ethyne or HC≡CH), these high-pressure reactions involve catalysis by heavy metal acetylides (commonly copper acetylide) to develop a large range of new synthetic routes for organic functional groups.

The pressures were so high for these reactions to occur that Reppe needed to develop his own stainless steel reaction vessels, known as Reppe glasses, to run the reactions.

Polymerization

Polymerization is an important process in the petrochemical industry. This process involves the linking of monomers (a single molecule that can bind to others) into polymers (all the joined monomers). This linking forms large-scale materials with a wide variety of properties and applications. These polymerization reactions are subdivided into two main growths: step growth and chain growth. Step-growth polymers are formed from stepwise reactions of the reactive functional groups found on certain monomers. If the polymer from a step growth releases water, the polymer is known as a condensation polymer; if not, they are known as addition polymers.

If instead of functional groups, polymerization occurs through reactive free radicals or ions, the reaction is commonly referred to as a chain-growth polymer. This encompasses many types of polymerization reactions with various reactive species (carbocations, carbanions, free radicals, and so on) that don't produce small molecules as a byproduct. Most interesting to catalysis is a polymerization technique discovered in the 1950s that can employ either a homogenous or heterogeneous catalyst, usually a titanium-based heterogeneous catalyst, called *Ziegler-Natta catalyst* after its codiscoverers, that's used to synthesize polymers from alkenes. Despite the prevalence of the heterogenous Ziegler-Natta catalysts, there are numerous zirconium and titanium homogenous catalysts found as metallocenes or "sandwich compounds" that serve the same purpose — a metallocene is a metal cation sandwiched between two flat five-membered carbon rings called cyclopentadienyl anions. Originally, this catalyst was used for the polymerization of ethylene into linear polymer chains using $TiCl_4$ and $Al(C_2H_5)_2Cl$. Now, the term encompasses a large range of catalysts ranging from metallocenes to rare Earth metals.

Oxidation and hydrogenation

Oxidation processes involve the addition of oxygen to compounds. This can be carried out by three typical routes: organometallic and redox chemistry of palladium, chain reactions of radicals initiated by Co and Mn, or the selective transfer of oxygen from an oxidizing agent or organic hydrogen peroxide.

Similarly, hydrogenation involves addition of hydrogen to an unsaturated moiety such as an olefin. This reaction can be catalyzed with most of the d-block elements, but Group 8 is especially important because it produces the most active catalytic systems. Hydrasilation of alkenes is similar to hydrogenation except that hydrogen and SiR_3 from a silane (R_3SiH) is added across the reactive double bond:

$$RCH{=}CH_2 + R_3SiH \rightarrow RCH_2CH_2SiR_3$$

Asymmetric molecules are important to living systems. For example, proteins are composed of amino acids that occur only in the levo form. As such, asymmetric catalysis has found wide-scale application in both the biochemical and pharmaceutical industries. Products with over 95 percent stereoselectivity can be achieved by fine tuning the environment where the ligand is bound to the catalyst. A tragic example of the importance of stereochemistry is the morning sickness drug, thalidomide. The (R)-enantiomer is effective in alleviating the symptoms of morning sickness; however, the (S)-enantiomer, the same molecule with a different 3-D structure, is a known teratogen, or molecule that can cause birth defects.

Table 16-1 summarizes all of these important reactions, plus a few others not mentioned.

Table 16-1	Reaction Summary		
Common Name	*Reactant*	*Product*	*Metal*
Carbonylation	1. Methanol and CO	1. Acetic acid	1. Rh or Co
	2. Methyl acetate and CO	2. Acetic anhydride	2. Rh
	3. Methyl acetylene, CO, methanol	3. Methyl methacrylate	3. Pd
Hydroformylation	Propylene, CO, H_2	Butyraldehyde	n– Rh or Co
Hydrocyanation	Butadiene and HCN	Adiponitrile	Ni
Hydrogenation	Alkene or aldehyde and H2	Alkane or alcohol	Rh or Co
Hydrosilylation	Alkene and R_3SiH	Tetraalkylsilane	Pt
Metathesis	Alkenes or dienes	Rearranged alkene(s) or dienes	Mo, Re, or Ru
Polymerization	Ethylene, propylene, etc.	Polymers	Ti or Zr with Al, also Cr
Epoxidation	Propylene	Propylene oxide	Mo
Wacker reaction	Ethylene and O_2	Acetaldehyde	Pd and Cu

Heterogeneous

Heterogeneous catalysis is, for the most part, a surface process. Due to the complexities in analyzing a surface reaction, working mechanisms for heterogeneous catalysts are not as well understood as those for homogenous catalysis. Surface reactivity is influenced by factors at the atomic surface as well as other factors from macroscale environment, making it very difficult to understand. Perhaps most famously, Wolfgang Pauli commented on the confusing nature of surface phenomena: "God created the volume of matter, the surface was created by the Devil."

Remember, catalytic reactions require the shuttling of electrons to help make and break bonds. For this reason, the surfaces of heterogeneous catalysts can be functionalized to be either acidic or basic depending on whether you require an electron donating surface (basic) or an electron withdrawing surface (acid). Aluminum is useful because it can be both acidic and basic, a state known as being *amphoteric*.

Additionally, metals are often used as the catalyst sites for the same reasons as the homogenous catalysts. In the case of heterogeneous catalysts, however, the metals are deposited on the surface of substrates, such as alumina or silica. These metals can be deposited as a fine layer, patterned using wet chemistry techniques, or simply be in the form of nanoparticles, clusters of metal atoms under 100 nm in diameter.

The principle of a heterogeneous catalyst works by adsorbing a reactant (known as an *adsorbate*) to the surface of the catalyst. This usually begins with a physisorption, which is a weak van der Waals interaction between the surface and adsorbate. To make the reaction possible, the adsorbate must get even closer to the catalyst, which occurs through chemisorption, a process that forms chemical bonds (ionic or covalent) between the surface atoms and the reactant. The key difference between physisorption and chemisorption is that the adsorbate and surface are unchanged in physisorption, as opposed to chemisorption, where the adsorbate undergoes a chemical reaction.

As shown in Figure 16-4, a dihydrogen molecule is weakly attracted to the surface of the catalyst (physisorption). For chemisorption to occur, the adsorbate chemically bonds to the surface of the substrate. This causes the bond between the hydrogens to break, making them free to undergo reactions with other nearby chemicals. When the final product is created (the free hydrogen atoms binds to another reactant), the final product undergoes desorption. Traditionally, this flows in the gas stream for collection at the end.

The *desorption*, or the molecule being released from the substrate, is often the slowest or rate limiting step. Because this controls the speed of your production, you must make careful consideration about the metal you choose and ensure that it doesn't bind too strongly to the product. A comparison of the binding affinity of the reactants and products with the metal surface is useful in catalyst selection.

Figure 16-4:
A hydrogen molecule is attracted to the surface of a catalyst with physisorption. When the hydrogen molecule begins to chemical react with the surface, it converts to a temporary transition state before finally being chemically adsorbed to the surface through chemisorption.

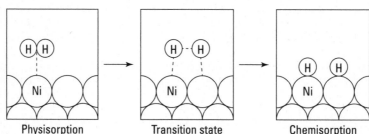

Physisorption   Transition state   Chemisorption

A good catalyst has only a slightly stable bond with the products. In many ways, it's like Goldilocks: It must be just strong enough to undergo chemisorption, but not too strong that it prevents desorption. In short, it must be *just* right.

For a simple example of heterogeneous catalysis, think of the catalytic converter in your car. It's made of a platinum-rhodium alloy that is finely deposited on aluminum oxide (specifically γ-alumina) to decompose nitrogen oxides to less toxic nitrogen, as well as promote the combination of O_2 with toxic CO and hydrocarbons to form CO_2.

In a typical reaction, just like the one occurring in your catalytic converter, the reactants enter as a gas, react with a metal surface, the catalyst makes the reactants undergo a change, and then products are collected (or released into the atmosphere) at the end of the gas stream. For this reason, heterogeneous catalysts are very desirable in industry because the product does not have to be separated from the catalyst.

Heterogeneous catalysts are often very porous materials or else have a very high surface area to increase the chances of interactions. An ordinary dense solid is not desirable because of the small surface area available for reactions. For example, α-alumina has a surface area of 5 m^2/g compared to γ-alumina, which has a surface area of 100 m^2/g; for this reason, γ-alumina is used more often for catalysis. This is also the reason that silica (SiO_2) is used instead of quartz.

Organocatalysts

Many important commercial products are created using organometallic compounds, or compounds containing bonds between carbon and a metal. This commercial interest is one of the main reasons for studying these types of compounds. As mentioned earlier in this chapter, their mechanism of action in homogenous catalysis is well-understood compared to heterogeneous catalysis.

Vinegar, or acetic acic, can be made simply through the fermentation of ethanol — yes, fermenting what was fermented! However, this process is relatively slow. Acetic acid can be produced much quicker using a rhodium complex that causes the carbonylation of methanol. This process, known as the Monsanto process (named after the Monsanto company), accounts for over a million tons of acetic acid a year. The catalytic cycle is a six-step process, as shown in Figure 16-5.

The catalyst is the anion *cis*-[Rh(CO)$_2$I$_2$]$^-$. The first step is to form [(CH$_3$) Rh(CO)$_2$I$_3$]$^-$ through the oxidative addition of methyl iodide. The methyl group on the Rh moves to the carbonyl ligand, changing the six-coordinate anion into a five-coordinate anion [(CH$_3$CO)Rh(CO)I$_3$]$^-$. This final complex can react with carbon monoxide to form a six-coordinate dicarbonyl complex that decomposes to form acetyl iodide (CH$_3$COI) and regenerate the starting catalyst anion. It's the CH$_3$COI that is finally hydrolyzed to make the final acetic acid.

Alkene polymerization is carried out using the Ziegler-Natta catalyst (shown in Figure 16-6), an extremely important process industrially. This process makes strong and rigid materials due to the isotactic polymers that are all branched on the same side and pack together with very high densities. The catalyst has a titanium center that adds an alkane, an organic compound with only carbons and hydrogen atoms with single bonds (known as alkylation). The new alkene molecule then binds on a neighboring vacant site. The alkene finally undergoes insertion to the Al-C bond creating another vacant site for another alkene molecule. This process is repeated, which results in long chains of alkanes.

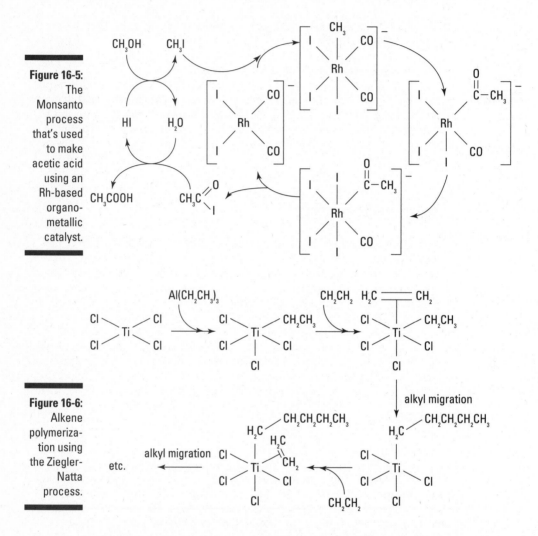

Figure 16-5:
The Monsanto process that's used to make acetic acid using an Rh-based organo-metallic catalyst.

Figure 16-6:
Alkene polymerization using the Ziegler-Natta process.

Hydrogenation of alkenes using Wilkinson's catalyst is one of the most studied catalytic systems (shown in Figure 16-7). It can hydrogenate a wide range of alkenes at atmospheric or reduced pressures. The cycle involves oxidative addition of hydrogen to a 16-electron rhodium(I), rendering an 18-electron rhodium(III) complex. A phosphine ligand is lost, making the molecule unsaturated. This unsaturated molecule then reacts with the alkene. The hydrogen then transfers to the alkene from the rhodium, followed by reductive elimination of the alkane.

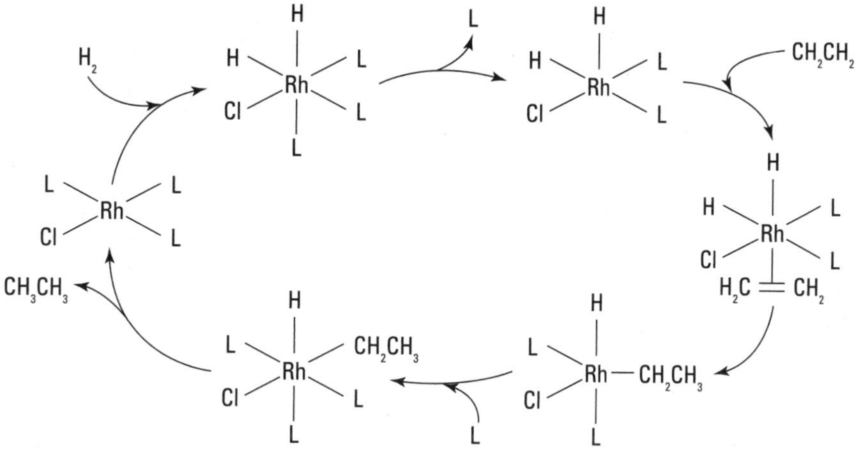

Figure 16-7:
The Wilkinson's catalyst is used for the hydro-genation of alkenes.

Chapter 17

Bioinorganic Chemistry: Finding Metals in Living Systems

*I*norganic molecules are found everywhere in the environment — even inside your body. For example, when your body absorbs certain vitamins and minerals, it creates large molecules that have a metal center. These *biomolecules*, created through biological chemical reactions and incorporating metals, are the basis of bioinorganic chemistry. *Bioinorganic chemistry* is the study of biological chemical reactions that involve an inorganic component.

In this chapter the bioinorganic chemistry of natural systems is described, such as photosynthesis, nitrogen fixation in plants, and how living creatures use metals (iron or copper) to transport oxygen to cells. There is also some explanation of environmental chemistry, including the study of toxins — both naturally occurring and synthetic (manmade). When humans and other organisms interact with pollutants in the environment, it's often through bioinorganic reactions that our bodies absorb these harmful molecules and elements. In the final section of this chapter we describe some of the most commonly used methods of detecting and measuring toxins and metals in the environment, as well as some commonly used techniques to remove contaminants.

Focusing on Photosynthesis

Photosynthesis is a chemical reaction that converts sunlight into energy for living organisms. It's one of the most remarkable reduction-oxidation reactions scientists have observed (see Chapter 3 for details on this type of reaction). There are two currently known systems of photosynthesis: Photosystem I (PSI) and Photosystem II (PSII). They each use the light in a series of reactions that pass on electrons. The essential difference is that PSI creates hydrogen sulfide (H_2S) by reduction of CO_2 and is only conducted by anaerobic bacteria and was more common in prehistoric times.

The PSII system will be the main focus because it's more important to the way that humans live and breathe. The PSII system creates oxygen gas (O_2) by reduction of carbon dioxide (CO_2) and is conducted by all green plants and algae.

During photosynthesis a molecule of carbon dioxide (CO_2) is reduced, and a molecule of water (H_2O) is oxidized. (See Chapter 3 for details on reduction-oxidation chemistry.) The result of these reactions is the production of dioxygen (O_2) gas and a complex *glucose*, or sugar molecule, as illustrated in the following equation:

$$6\,CO_2 + 6\,H_2O \rightarrow C_6H_{12}O_6 + 6\,O_2$$

Energy to fuel this reaction is captured from sunlight and transformed through a series of reactions using *chlorophyll*. Chlorophyll is a pigment, or light absorbing molecule in the plant. The absorbed light energy is converted to chemical energy that is then used to break down CO_2, making O_2 and sugar in the process. Chlorophyll is found in plant cell organelles called *chloroplasts*, they absorb light from two parts of the electromagnetic spectrum — the red part, and the blue part. The color of the chlorophyll pigment (green) comes from the light that's reflected, not absorbed. It's reflected and then observed by your eyes.

Chlorophyll complexes contain *porphoryin* molecules. These large ring structures have four nitrogen atoms that donate electrons to a metal center and help bind it. For example, take *chlorophyll a*, illustrated in Figure 17-1.

Chlorophyll a is a *magnesium dihydroporphyrin complex*. Because the magnesium (Mg) isn't on the same plane, *coplanar*, with the nitrogen (N), the bonds are represented by dashed and straight lines in Figure 17-1. If you looked at the ring structure from the side, you would see the magnesium atom located about 30 to 50 pm above the nitrogen atoms. Usually, the magnesium is also bonded to another *ligand*, or connector, to keep it in place. In the case of chlorophyll a, the other ligand is usually a water molecule.

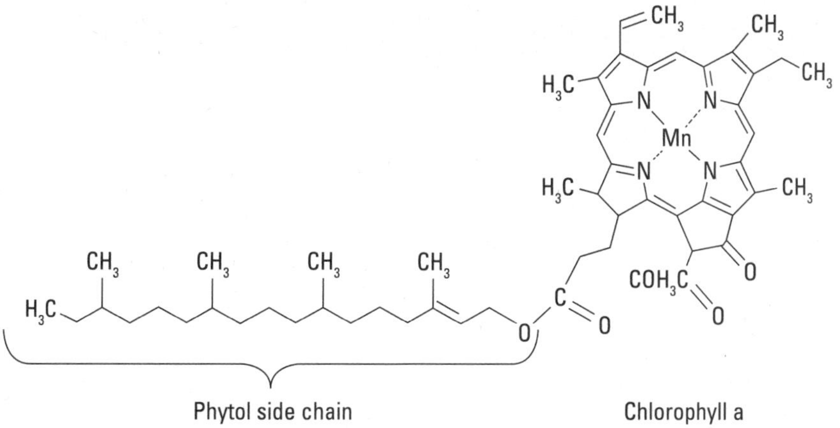

Figure 17-1:
Photosyn-
thesis in
leafy green
plants is
carried out
by a mol-
ecule such
as chloro-
phyll a.

Phytol side chain

Chlorophyll a

The ring structure of the porphyrin molecule is what absorbs the light. It has alternating single and double bonds that interact with the electromagnetic radiation coming from the sun. The energy absorbed by the ring structure excites the electrons enough that they move from π to π^* orbitals. These more energetic electrons are needed to break apart the CO_2. Actually, the decomposition of carbon dioxide isn't thermodynamically favorable, it requires a large amount of energy. In fact, CO_2 is so stable that when formed, it's said to be the death of carbon, because that carbon is very difficult to use thereafter. This serves as a testament to the ingenuity of mother nature, that it has found a way to use the CO_2 using an abundant energy source, the sun.

The oxygen that you breathe comes from the water molecules involved in photosynthesis, not from the carbon dioxide, as is commonly thought. The exact mechanism for the release of oxygen gas (O_2) is still unclear, but scientists have observed that it occurs in a four-step process. This has been tested by shining light on plants in pulses and observing that after the fourth pulse of light some O_2 was released. The reaction is summarized by the following equation:

$$H_2O \rightarrow O_2 + 4H^+ + 4e^-$$

The four electrons are shuffled around the porphyrin, creating a series of oxidation and reduction reactions. During photosynthesis the PSII molecules become oxidized, and they must be reduced before they can operate again. The energy required to convert PSII from the oxidized to the reduced form is used to drive the conversion of two biomolecules of ADP (adenosine diphosphate). ADP is converted into two molecules of ATP (adenosine triphosphate), which is then used as a source of energy in the cells. Yet another example of the ingenuity of Mother Nature. In one step the PSII system is regenerated, and ATP gets charged up so it can give energy back within the cells.

Climbing Aboard the Oxygen Transport

After oxygen gas has been released by plants through photosynthesis, it's absorbed by other living creatures, including humans who inhale it from the air. When in your lungs, however, the oxygen must be transported to all of the cells in your body. That's where metals — and thus inorganic chemistry — play another important role.

Three types of proteins have evolved to transport oxygen throughout the body.

- ✔ **Hemoglobin:** Hemoglobin has an iron atom at its center and is present in red blood cells of vertebrates (organisms with a backbone, such as humans).

- ✔ **Hemocyanin:** Hemocyanin has a copper atom in its center and is found in mollusks and arthropods (organisms such as snails and insects).

- ✔ **Hemerythrin:** Hemerythrin also has an iron atom in its center like hemoglobin, but hemerythrin is found only in certain species of sea worm.

Hemoglobin proteins comprise 65 percent of the iron (Fe) in the human body. Hemoglobin is a flat planar-shaped molecule that's springy and has a heavy magnet (the iron atom) in the middle. In hemoglobin, iron is coordinated by four nitrogen atoms.

The iron at the center of the hemoglobin (Figure 17-2) acts like a snare, capturing oxygen. When dioxygen gas (O_2) is added to the iron, the neighboring ring atoms cooperate to pull the O_2 molecule closer. When the iron binds to the O_2, it becomes temporarily oxidized to Fe^{3+}, and the dioxygen molecule becomes O_2^-. The effect of adding O_2 causes a shift in the energy of Fe^{2+}, moving it from a high-spin state to a low-spin state. (See Chapter 13 for further details about spin states.)

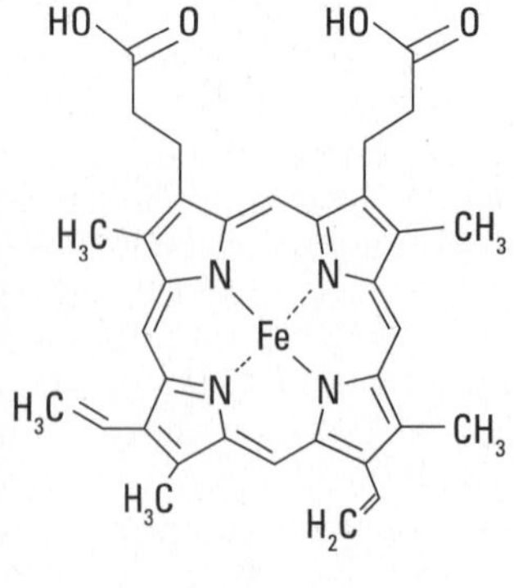

Figure 17-2: Hemoglobin binds to dioxygen molecule using iron.

When the iron changes spin states, the ion becomes smaller. This allows it to be pulled in even closer. As it moves into the center of the ring structure, it causes the porphyrin ring to flatten out. The result of this interaction is that the O_2 is pulled into the protein by the iron, but can be released easily, too.

The dioxygen (O_2) molecule serves as a π acceptor when interacting with the Fe^{2+} center of the protein porphyrin structure; therefore, other π acceptor molecules such as nitric oxide (NO), carbon monoxide (CO), cyanide (CN^-), isocyanide (R-NC), azide ion (N_3^-), and thiocyanate (SCN^-) can stop hemoglobin from properly capturing and transporting oxygen; ultimately leading to asphyxiation and death.

To avoid bonding with other, possibly harmful molecules, several histidine molecules — an amino acid — are placed around the oxygen-binding location, that effectively guards the location. There are two histidine molecules attached to each heme group, or subunit, in a hemoglobin complex. These cause a distortion in the protein shape so that only a molecule with a nonlinear geometry can bind to it. Similar to puzzle pieces, only the atom with the correct shape and symmetry can fit comfortably.

Feeding a Nitrogen Fixation

All life requires nitrogen to build compounds such as proteins and nucleic acids. The air you breathe is 79 percent nitrogen, but it's triply bonded dinitrogen (N_2) and very hard to separate. N_2 is an almost inert molecule, the triple-bond has an energy 945 kJ/mol, this means that a large amount of energy is required to break apart the bonds to separate the nitrogen atoms so that they can be used in living systems.

Even though nitrogen is the most abundant element in the atmosphere, most of it can't be used by organisms as it is quite inert. Before nitrogen can be used by an organism, it must first be fixed, or broken apart. *Nitrogen fixation* is accomplished in multiple ways and is one part of the *nitrogen cycle*.

The *nitrogen cycle* is one of many important nutrient cycles on earth (the others include the carbon cycle, sulfur cycle, and phosphorus cycle). The nitrogen cycle transfers nitrogen through the environment — from the air to living organisms and back to air again through a series of chemical reactions.

Fixing nitrogen for use by organisms

For plants and animals to access the abundance of nitrogen in the air, it must first be converted to a useable form. The process of *nitrogen fixation* that breaks apart the N_2 molecule is summarized by this formula:

$$N_2 + 6e^- + 8H^+ \rightarrow 2NH_4^+ \text{ (ammonium ion)}$$

This process can happen in three different ways: by lightning in the atmosphere, by bacteria in the soil, or by manmade processes.

Atmospheric fixation occurs when lightning supplies the energy to break apart the N_2 molecule, forming nitric oxide molecules:

$$N_2 + O_2 \rightarrow 2NO$$

The nitric oxide that is produced further combines with oxygen to form *nitrogen dioxide* (NO_2):

$$2NO_2 + O_2 \rightarrow 2NO_2$$

The nitrogen dioxide dissolves in water to produce nitric acid (HNO_3) and nitrous acid (HNO_2):

$$2NO_2 + H_2O \rightarrow HNO_3 + HNO_2$$

These acids easily release hydrogen, (a characteristic that gives them some explosive power). Upon release of the hydrogen, nitrate and nitrite ions are formed, which can be absorbed by plants.

Biological fixation is done by specialized microbes that convert the N_2 into ammonia (NH_3^-), nitrate ions (NO_3^-), or urea ($(NH_2)_2CO$). Although ammonia can be taken up by plants through the roots, it's more often the ammonium ion (NH_4^+), that's the conjugate acid of ammonia, that is used by biological systems. (See Chapter 5 for a quick recap on conjugate acids and bases.)

Several bacteria, blue-green algae, and yeasts can fix nitrogen. For example, *rhizobium* is a microbe that lives in root nodules of legume plants (clover, alfalfa, beans, peas) and are some of the most important nitrogen-fixing species of bacteria. (This is why it's great to plant these in your vegetable garden as off-season cover crops, because they add nitrogen to the soil that the next cycle of crops can use.)

Within nitrogen-fixing bacteria, an enzyme (composed of proteins) called *nitrogenase* is responsible for nitrogen fixation. Within both of these bacteria the active site for N_2 binding involves a molybdenum atom.

Industrial fixation occurs through reactions such as the *Haber-Bosch reaction.* In the Haber-Bosch reaction, nitrogen gas is passed over an iron catalyst at high temperature and pressures. Under these conditions the nitrogen breaks down; when in the presence of hydrogen, nitrogen bonds to it and forms ammonia. (See Chapter 16 for details about catalysis.)

When the industrial-scale process was developed to make ammonia, it boosted agriculture output greatly. Farmers could fertilize crops on a much larger scale than ever before. But the man-made production of ammonia for agricultural use has also affected the natural cycling of nitrogen through the environment, creating an overload of nitrogen that disrupts some ecosystems.

Re-absorbing nitrogen

When plants or animals die, the nitrogen in their bodies is still in its organic form. This nitrogen doesn't immediately convert back to N_2 gas and re-enter the atmosphere. As the organism decays, *ammonifying bacteria* work to transform the nitrogen (N_2) into ammonia (NH_3), a process called *ammonification.* Plant roots can then reabsorb the ammonia and put the nitrogen back to work building organic compounds.

Another way that nitrogen can be made available for plants is by *nitrification.* Nitrification is the process of converting nitrates (NO_3^-) and nitrites (NO_2^-) (released by decaying organic matter) into ammonium ions in the soil. The first reaction — oxidizing nitrogen gas to form nitrites (NO_2^-) — is completed by a specific genus of bacteria called *Nitrosomanas.* Following that, a genus of bacteria called *Nitrobacter* oxidizes the nitrites (NO_2^-) to nitrates (NO_3^-). This process includes some complex redox reactions but can be summarized by the following equation (remember that a redox reaction in practice is an electron-transfer process):

$$NO_3^- + 2e^- + 2H^+ \rightarrow NO_2^- + H_2O$$
$$NO_2^- + 6e^- + 2H^+ \rightarrow NH_4^+ + 2H_2O$$

Sending nitrogen back to the atmosphere

Whereas the processes described so far work to transform nitrogen into an organic form, there's also a process that works in the opposite direction, transforming nitrogen back into N_2 that's released into the atmosphere. *Denitrifying bacteria* convert ammonia as well as nitrate (NO_3^-) and nitrite (NO_2^-) back to gaseous nitrogen N_2. The dinitrogen gas is then released into the atmosphere, thus completing the cycle of nitrogen.

The nitrogen cycle used to be a closed-loop system, but recently the input has begun to exceed the output, therefore tipping the balance of nitrogen in the environment; in some ecosystems, the effect of this can be clearly felt. For example, when there is an over-abundance of nitrogen in an aquatic ecosystem (such as a lake or pond), more than the denitrifying microbes can deal with, it causes *algae blooms* to form. These arise when algae populations grow out of control, because they feed from the excess nitrogen that's available. The result is that they become overcrowded and block sunlight to plants and animals below the water surface. This effectively starves the ecosystem of much needed sunlight and causes plants and animals to suffer. Furthermore, when the algae blooms eventually die off and decay, they also use up all the dissolved oxygen within the water. This causes a case of oxygen deficiency in the water, and we all know that oxygen is necessary to sustain life. In the case of severe algae blooms, hydrogen peroxide (H_2O_2) has been used to oxidize and decompose algae and oxygenate the water at the same time. Hydrogen peroxide is used because as a catenated molecule, HOOH, it readily decomposes to H_2O with the evolution of OH^- that readily oxygenates the surrounding environment.

Being Human

The most important elements to life are **c**arbon, **h**ydrogen, **o**xygen, and **n**itrogen — the CHON atoms — as well as sulfur and phosphorus. These elements are combined in hundreds of ways to construct the building blocks of life, such as carbohydrates, fats, proteins, and nucleic acids.

But life can't be built on CHON alone. Living organisms also need metals. Major quantities of potassium (K), magnesium (Mg), sodium (Na), calcium (Ca), and chlorine (Cl) are found as ions in the body, or bonded with biomolecules. These Group 1 and Group 2 elements are commonly found in *isotonic* drinks. Their importance stems from the high-charge density they possess. Elements in both of these groups are eager to give up electrons, which can be donated to other biomolecules and processes within the body.

Sodium, for example, helps maintain osmotic pressure within cells to keep them from collapsing. Potassium, in turn, helps maintain correct levels of sodium in the cells by acting as an ion pump. (In fact, the energy created from pumping sodium in and out of the brain cells can amount to 10 watts of power output, which is almost enough to run a small light bulb.) Sodium is also responsible for filtering contaminants from the kidneys, generating the electric signal of your heart rhythm, and controlling how the eye focuses.

The d-block elements such as iron (Fe), zinc (Zn), and copper (Cu) are needed in smaller quantities and are found as *metallo-enzymes*, such as hemoglobin in your blood that transports oxygen to your cells (see "Climbing Aboard the

Oxygen Transport" earlier in this chapter), and others that are responsible for different electron-transfer and acid-base reactions in the body.

Consider the statement, "You are what you eat." You are what your food eats, too. For example, plants convert water and soil into fruits and vegetables that contain the minerals and elements from where they grow. These minerals and elements then pass along the food chain to us. Essentially, we eat the matter that makes up our environment. By virtue of these plants our bodies are built from the same elements that are found in the soil and water. We are of this planet, and not surprisingly, we share a similar chemical composition with both the Earth's crust and the oceans, but especially the oceans — from whence we came. This can clearly be seen in Table 17-1.

Table 17-1	Elements in Humans, Oceans, and Earth's Crust in Order of Abundance (in Descending Order)	
Humans	**Oceans**	**Earth's Crust**
H	O	O
O	H	Si
C	Cl	Al
N	Na	Fe
Na	Mg	Ca
K	S	Mg
Ca	Ca	Na
Mg	K	K
P	C	Ti
S	Br	H
Cl	B	P

Making things happen: Enzymes

Biological catalysts are known as *enzymes*. They are typically very large molecules that are composed of somewhere between 100 and 10,000 amino acids strung together in very specific sequences. They act within the body to rapidly carry out important chemical reactions. About one third of the enzymes in the body contain metals, and often the metal is the key component on which the reaction depends. Enzymes that contain metals are known as *metalloenzymes*.

Each enzyme has a specific function and purpose in the body, such as digesting food, transferring energy, and replicating DNA. Enzymes have a specific reaction center, known as a *substrate*. The substrate is of a very specific size, shape, and charge to ensure that only specific molecules are allowed to fit and operate correctly. It's similar to a lock and key, where the enzyme is the lock that can be opened only by the right biomolecule key for the substrate.

Enzymes are prevented from working properly if some other molecule interacts with the substrate instead of the correct biomolecules. When this happens the enzyme shuts down, causing the body to suffer in some way because it's prevented from working properly.

Here is a list of some important metalloenzymes in the body:

- ✔ **Oxidase:** Oxidase uses O_2 as an electron acceptor that is reduced to either water (H_2O), or to hydrogen peroxide (H_2O_2). In doing this, it converts the oxygen into a form that's more useable in the body.

- ✔ **Carbonic anhydrase:** Carbonic anhydrase acts as a buffer to maintain the pH of blood (see Chapter 5 for details on acid-base chemistry). It's built with zinc, which is one of the most important metals found in enzymes, and it's responsible for the proper functioning of over 3,000 enzymes, in all catalytic sites. The zinc ion functions as a Lewis acid; it's often found as Zn^{2+}. In carbonic anhydrase, the zinc is surrounded by three histidine groups (h).

$$h_3ZnOH_2 + H_2O \rightarrow H_3O^+ + h_3ZnOH^-$$

$$h_3ZnOH^- + CO_2 \rightarrow H_3O^- + h_3Zn$$

A buffer is a specific molecule that can prevent sharp changes in pH from occurring, therefore buffering the environment from the consequences of such changes.

- ✔ **Hydrogenase:** The hydrogenase enzyme is responsible for the reduction and oxidation of hydrogen (H_2) in the body, described by this formula:

$$2H^+ + 2e^- \rightarrow H_2$$

Hydrogen plays an important role in biochemistry, so having access to the oxidized or reduced version allows many biochemical reactions to occur as needed. More importantly, hydrogen is an electron acceptor (instead of oxygen) for anaerobic metabolism. Enzymes play an important role in helping digest polymers (large, complex molecules) of food. Most digestive enzymes act as catalysts to speed up the hydrolysis reaction. Here are two of the important digestive enzymes:

  - **Pepsin:** Pepsin is released in the stomach and breaks proteins down into smaller units, eventually getting down to the amino acids. It does this at a pH of 2. In the digestive system the pepsins are classified according to 1 of 12 types of amino acid, which they are responsible for converting.

- **Pepsinogen:** In the stomach, hydrochloric acid is released and creates an acidic environment (pH 2). At this low pH, pepsinogen enzymes convert to pepsin. Whereas pepsinogen may be present elsewhere in your body, it can't convert to pepsin and begin to digest proteins until it encounters the highly acidic stomach fluids. This keeps the pepsinogen from digesting proteins in parts of your body where it shouldn't.

At the very core of human existence, written in an alphabetic code of three letters, you find *deoxyribonucleic acid*, or DNA. The DNA in your cells holds all the genetic information that makes you *you*. And this is where metalloenzymes play an important role. They unravel the double helix strand of DNA molecules; they read each strand; they make copies, and even correct mistakes along the way.

Curing disease: Medicines

Unlike the large and complex enzymes listed previously, medicinal molecules are often much simpler, consisting of ten or so combined molecules. They typically have a metal center, using elements such as lithium (Li), platinum (Pt), gold (Au), silver (Ag), and bismuth (Bi).

Lithium is used as a mood stabilizing drug. The Li^+ ion interacts with many nerve centers throughout the body and within the brain. Li^+ has a high-charge density and gives up its electrons readily so it can interact with a variety of biomolecules. Scientists aren't certain how lithium functions to stabilize mood, but it's been used since the 19th century nonetheless.

Gold is, for the most part, a very inert metal, and for this reason it's been used in jewelry for millennia. However, in recent years with the advent of nanotechnology (see Chapter 19), it's been found that size-dependent effects occur for particles of gold that are in the size scale less than 100 nm in diameter. These nanoparticles interact with light in very specific ways. Importantly, they can interact with a laser light that won't interact with skin and muscle tissue of the body. As such, the laser can cause the nanoparticles to heat up, and when attached to cells such as cancer, they can destroy the cancer. This is an emerging technology, but hopefully soon cancer will be a thing of the past.

Silver has been used for millennia because of the sanitation properties, to the extent that it's written into our language — ever hear of someone who was fed with a silver spoon? This comes from the fact that silver is an expensive metal that's known to kill pathogens. During the black death, only the rich could afford to eat without fear of catching the plague from their cutlery.

Today, silver nitrate $AgNO_3$ is used to cauterize wounds by converting nitrate to nitric acid. At the same time acting as a sterilizing agent and preventing infection with its anti-microbial properties. Silver is sometimes used to purify municipal water supplies.

Everyone has probably used some Pepto-Bismol at one time or other (this is more common in the United States than in the other parts of the world). The active ingredient is based on a bismuth metal center. It is taken to alleviate upset stomachs. If you consume large quantities your tongue turns black due to the formation of bismuth sulfide — it binds to the sulfur from proteins in the body (and on the tongue); bismuth sulfide is a black material.

Cisplatin is a cancer-fighting drug that's a square planar platinum molecule (see Chapter 9). It's a relatively small molecule that binds to the backbone of a cancer cell's DNA. It kinks the structure of the DNA strand and prevents it from operating. The cisplatin binds to the DNA at two locations; in effect, it acts as a multidentate ligand, or chelating complex (see Chapters 9 and 15 for more details). The chelating effect changes the symmetry of the DNA molecule. It's like cisplatin reaches out with two hands and pulls the DNA in such a way that it can't operate properly anymore.

Causing problems: Toxicity

The softness of a heavy metal indicates how toxic it is. (In this case, *softness* means a heavy metal ion with a low charge.) Soft, and therefore toxic, metals include mercury (Hg), lead (Pb), cadmium (Cd), and thallium (Tl). Mercury and lead in particular are strongly complexing and, therefore, very toxic in the body because they can bond in biomolecules and prevent them from working properly, as well as be difficult to remove. (See Chapter 5 for a refresher on hard and soft acids.)

Often, toxicity arises due to *competitive binding* in an enzyme by another molecule. The competing molecule, known as an *inhibitor*, may fill the reaction space meant for an enzyme. When this occurs, it's called *competitive inhibition*. In other cases, toxicity may occur because the metal bonds somewhere with the enzyme and distorts its shape and/or charge so that it can't successfully function as it's supposed to. This is known as *noncompetitive inhibition.*

Metals can be especially toxic if they have become *methylated*, this can be done by anaerobic bacteria. A methyl group $-CH_3$ can make the metal more biocompatible, so it can move through the blood stream and thus all around the body, and more importantly, it can be passed up along the food chain. For example, mercury can find its way into the ecosystem by way of industrial output. It then ends up in water supplies and sinks to the water bed where it encounters anaerobic bacteria that form methylmercury $[CH_3Hg]^+$.

This can bond readily into the materials of which plants are made, that are then consumed by animals, and ever larger animals that prey on those. The methylmercury moves up the food chain, and eventually it can end up on your dinner plate.

Other metals that can be toxic include sulfur (S) and fluorine (F). Here are some examples of how they work:

- ✔ **Sulfur:** Hydrogen sulfide (H_2S) is a product of photosynthesis by ancient microbes. For this reason, sulfur is found extensively within oil and gas deposits (oil and gas are the decomposed remains of ancient life). So it becomes released into the atmosphere when natural gas is extracted. Sulfur smells bad; it is the smell of rotten eggs. H_2S has the effect of knocking out your sense of smell at lethal concentrations, so as long as you can smell it, you're in good shape! Fortunately it's strong enough that it can be detected at very low concentrations. The average concentration in the air is 0.05 ppb.

- ✔ **Fluorine:** Hydrofluoric acid (HF) is a by-product of aluminum production by electrolysis and the synthesis of chlorofluorocarbons (CFCs). Although in water HF is a weak acid, it's extremely toxic. HF poisoning causes deep ulceration or scarring of body tissue with a delayed effect, slowly replacing calcium in bodily tissue (flesh and bones). There's no pain at first because it's a weak acid, but when the calcium in the bones starts to be replaced by the more electronegative fluorine it feels similar to growing pains, and may be too late. Merely five-minute exposure to 10,000 ppm concentration in the air is fatal to humans.

Answering When Nature Calls: Environmental Chemistry

Environmental chemistry deals with the chemistry of natural systems. This includes the study of geochemistry, metal deposits, environmental pollutants, "green chemistry," and hazardous materials. In the environment, many elements can be considered contaminants, from manmade organic pesticides to heavy metals such as mercury and arsenic. Some chemicals occur naturally in the environment, due to mineral deposits or volcanic gasses. Yet others enter the environment from manmade sources such as fertilizers, industrial pollutants, or even from prescription drugs in the body that pass into the water supply systems.

In this section a simple description of how scientists recognize when contaminants are in the environment is presented, as well as some of the most commonly studied environmental pollutants, including pesticides and heavy metals.

Eyeing key indicators

The environment is a large, complex, and interrelated system where a change here or there can have unknown consequences somewhere else. Despite the complexity, scientists try to understand the environmental system and even predict what will happen next (weather reporting is an example of how scientists try, and sometimes fail, to accurately predict how the environment will behave in the future). To understand the system and monitor it for patterns and change, scientists look to *key indicators*.

Key indicators can be physical (sea surface temperature), chemical (oceanic CO_2 concentrations), or biological (algae bloom). Indicators help scientists understand environmental systems, and in particular, help scientists identify when an environmental system is in danger, or imbalanced.

Useful characteristics for environmental indicators include the following:

✔ An indicator should be responsive, or sensitive to changes in the environment, both natural and manmade.

✔ An indicator should have a sound theoretical basis that can be measured in technical and scientific terms.

✔ An indicator should be based on international standards where there is international consensus about its validity.

✔ An indicator should be comparable to international settings so that international comparison can be made. A change in one region often affects other parts of the world. For example, the Chernobyl nuclear disaster changed the air quality all over northern Europe and was even detected on the far edge of Europe at the Mace Head Atmospheric Research Station in County Galway, Ireland — over 3500 Km (2200 miles) away!

✔ Indicator data must be quantifiable so that it can be measured, tested analytically, and compared to other values.

Keep reading to find out how scientists use oxygen levels and pH as key indicators of environmental change.

Reading oxygen indicators

Earth is an oxygen-rich planet with a large amount of water vapor (H_2O) in the atmosphere. Oxygen is vital to the living beings that populate Earth, including humans. Oxygen compounds can be found almost everywhere and in some cases can be excellent indicators of environmental change or pollutants.

There's an important layer of air in the upper atmosphere (or *stratosphere*) composed of ozone (O_3 molecules) called the *ozone layer*. The ozone layer protects life on the earth's surface (including you) by blocking harmful ultraviolet (UV) rays from the sun. Molecules of ozone absorb part of the UV light spectrum radiating from the sun and prevent it from burning the earth's organisms.

Scientists observing the ozone layer using spectroscopy (see Chapter 21), high-altitude weather balloons, and earth observation satellites determined that the ozone layer was deteriorating. In particular, an area of the ozone layer above the Antarctic had become much thinner over time, so thin that it is now commonly referred to as the "hole in the ozone layer." Scientists soon realized that ozone can be depleted by organic molecules such as chlorofluorocarbons (CFCs) that used to be common in fridges, gas canisters, and aerosol cans. Based on this assumption the global use of CFCs was greatly reduced.

Oxygen is also used to indicate the presence of pollutants in water. The levels of *dissolved oxygen* in water decreases dramatically when excess organic matter (often from wastewater) enters the ecosystem. The organic matter is oxidized by microorganisms. This effectively takes the oxygen in the water and bonds it to the organic matter. This noticeably reduces the dissolved oxygen in the system. The absence of dissolved oxygen has detrimental effects on fish and other organisms. Scientists *oxygenate* the water using hydrogen peroxide to improve conditions.

There are several oxygen compounds that are caustic to the environment. Carbon monoxide is a toxic gas that forms upon the combustion of hydrocarbons, such as the gasoline you use in a car. Catalytic converters are needed to break down these poisonous compounds (see Chapter 13 for more details). Carbon monoxide is produced by the combustion reaction:

$$2C_8H_8 + 17O_2 = 16CO + 18H_2O$$

Fog and smog can be indicators of air quality, but they are different because of the origin of the materials that onset droplet formation, which in turn form a dense cloud in the atmosphere. A particle acts as a *nucleant* by providing a surface area upon which water droplets can adsorb and condense. When water condenses onto something natural, such as pollen, it's called *fog*. When it condenses on something that is manmade or synthetic, it's referred to as *smog*. (Smog can also be caused by smoke as a result of large ground fires.)

Changing pH

Throughout the environment, scientists observe organisms (including humans) that are sensitive to changes in pH.

For example, lichen is a plant-like organism that grows in some of earth's most extreme environments. Many lichen species are highly pH sensitive, thriving only in a narrow range of pH conditions. A minute change in pH can cause lichen to change color or stop growing. Changes to lichen color were noticed on a grand scale in the Black Forest in Germany, indicating to scientists that something was changing the environmental pH. This was later attributed to a condition known commonly as acid rain.

Another example of pH indicating the presence of contaminants in the environment is the *acid rain* phenomenon. Due to industrial air pollution containing sulfur dioxide (SO_2), atmospheric water became more acidic. Scientists first noticed this change in rain water when they observed ancient buildings such as monasteries and castles being damaged by this acid rain. Soon after, they realized sensitive ecosystems such as the Black Forest in Germany were also being affected. For this reason the pH of rain water is recorded so that scientists can continue to observe differences over time and understand how pollutants that enter the atmosphere affect the environment.

The oceans and seas are other places where changing pH levels have caused changes in the environment. Due to modern industrialization, large amounts of carbon dioxide (CO_2) have entered the atmosphere. CO_2 is absorbed into the oceans and changes the pH, making the waters more acidic. Scientists are currently studying how these changes in pH affect the organisms that live in the oceans.

Rocking the heavy metals

Heavy metals are not found widely distributed or very common in nature, never mind in our own bodies.

Heavy metals often enter the environment from industrial processes, so the concentration of the these contents in products and a as result of industrial production can be monitored to determine human risk and safety factors.

Some of the most common (and dangerous) heavy metals include:

- ✔ **Mercury:** Mercury is a strong complexing agent, but it is most dangerous methylated. Mercury is the inspiration for the Mad Hatter from Alice in Wonderland. Hatters commonly employed elemental mercury in the process of hat making. By inadvertently inhaling the vapors, over time these hatters went mad due to the continued absorption of mercury.

- ✔ **Lead:** Lead is a common material found in the effluent of several industries. In particular, the form tetraethyl lead ($(C_2H_5)_4Pb$) was added to gasoline and thus made its way into the air and soil, after it is ejected from a combustion engine. Fortunately it's not used as an additive anymore. However, lead is also found as a pigment in paint and can be consumed as paint dust or paint chips. This is still common to find today.

Some lead pigments and their respective colors include: lead carbonate $2PbCO_3 \cdot Pb(OH)_2$ (white), lead chromate $PbCrO_4$ (orange, red, and yellow), and lead oxide PbO (scarlet, red).

✔ **Arsenic:** Arsenic is a heavy metal contaminant that occurs in drinking water globally. Efforts are under way to implement iron oxide nanoparticles that have a strong binding affinity with arsenic. Arsenic is a large molecule and hard to bond with, but based on the size of the nanoparticle and the oxidation state of the iron, it can adsorb the arsenic onto the nanoparticle surface, and in the presence of a magnet the magnetic nanoparticles are extracted. This in turn carries away the dangerous arsenic metal from the water supply, too. For more information see Chapters 13 and 19 for more details about magnetic nanoparticles, and Chapter 9 for details about magnetism.

Killing me softly: Pesticides

Pesticides are chemical compounds used to control or destroy pests such as weeds, insects, and rodents (just to name a few).There are two types of pesticide: *organic* (compounds that are predominantly composed of carbon) and *inorganic* (compounds with some metal involved). Whereas inorganic pesticides are the oldest known, these days the majority of pesticides are organic and are used on a large scale.

Inorganic pesticides are of mineral or metallic origin. These include borates, silicates, zinc, and sulfur. These materials are mined and then ground into a fine powder. Others that were previously popular include arsenic, copper, lead, fluorine, and tin salts. Some inorganic pesticides are bonded with organic ligands, such as the case with carbonyl chloride $COCl_2$ (also known as phosgene) or cyanide (HCN triazine pesticide). Some commonly used inorganic pesticides include:

✔ **Bordeaux mixture:** This gets its name from its early use to protect grapevines in the wine region of Bordeaux, France. Specifically it's used to kill fungal material from plants such as fruit trees and vines. It's made of equal portions copper sulfate ($CuSO_4$) and calcium oxide (CaO) or slaked lime ($Ca(OH)_2$).

✔ **Borate insecticide:** This is boric acid ($B(OH)_3$) and is sold as Bora care and Timbor. It's used to protect wood against damage from insects.

✔ **Sulfur and lime-sulfur (calcium polysulfide):** These are the oldest-known pesticides. It's used in horticulture of deciduous trees, particularly Bonsai trees. Lime-sulfur is used to kill off fungi, bacteria, and insects. It has the general formula CaSx, where x is an integer value.

Organic pesticides can be found naturally occurring (such as nicotine), or they can be made by microorganisms such as bacteria, fungi, or viruses. These toxic carbon compounds can also be industrially synthesized (organo-phosphates, for example).

Looking for and removing contaminants

Traditional methods of detection rely on three basic analysis techniques: volumetric, gravimetric, and combustion analysis. In all three you expect a certain chemical reaction to occur, such as a blue-colored solution to turn red when you reach a certain pH or concentration. When adding liquids to each other you are performing a *volumetric* reaction. If you create a product such as a salt that might precipitate from the solution, then you can weigh the material to complete a *gravimetric* measurement. Or you could watch for the complete oxidation of elements such as C, H, N, and sometimes S. These would undergo a *combustion reaction* and form CO_2, H_2O, N_2, and SO_2. Using one of these three techniques or many of them at the same time, you can get a clear understanding of the materials that are present in a sample.

Toxins in the environment occur naturally, such as mineral deposits that make rivers flow red, or they occur as by-products of industrial processes. In either case, scientists have found ways to look for and identify contaminants in the environment. Here are some of the most common ways scientists sample for and study the level of contaminants in the environment:

✔ **Flotation:** This method exploits the difference in the surface properties of materials in a fluid system. It can be used to remove solid particles, liquid droplets, chemicals, ions, and even biological compounds. It's used in the mining industry for the processing of secondary materials, to treat materials from water run-off that may arise from treatment and washing stations, thereby preventing the run-off from entering local water supplies. And it can be used to separate ions from effluent treatment stations. Flotation is a cheap and relatively efficient technique.

✔ **Soxhlet extraction:** Also called liquid-solid extraction, this is a common tool and is sometimes used as a standard to compare new techniques and methods of extraction. It works by the continuous cycling of a boiling solvent, whereby over time the remaining solid materials in the solvent collect together and can be removed. It's most often used to remove organic pollutants from air and water.

✔ **Supercritical fluid extraction:** This can be any compound that at a certain temperature and pressure is above the critical point. The *critical point* of a material is the place on a phase diagram where solid-liquid-gas intersect. For water the critical point is 4 °C (39.3 F). But when the material is above the critical temperature, no amount of pressure can make it liquefy. When in a supercritical state, only one phase can exist; but it has interesting properties in that it can have solvent properties of a liquid, but have transport properties of a gas. The extraction process works because the solubility of the compound to be extracted (solute) depends on temperature and pressure. Under the appropriate conditions the solute can then be extracted. This technique is used to extract materials such as caffeine, nicotine, and essential oils.

✔ **Water purification:** Water filters that are commonly found in grocery stores are designed to extract contaminants from drinking water. Such contaminants can include heavy metals, minerals, and prescription drugs that are not extracted by municipal water processing plants. They work in two stages that employ an ion-exchange resin and activated carbon in unison. The ion-exchange resin can bind minerals and neutralize the anions and cations on the resin surface. Activated carbon is porous and is responsible for binding organic contaminants that include bacteria, for example. Water filters must be replaced periodically because there is no means by which the bacteria are destroyed. If the filter is not replaced, it becomes saturated and rendered ineffective. Basically, the filter becomes so dirty that there's no more room to absorb any more contaminants.

Chapter 18

Living in a Materials World: Solid-State Chemistry

. . . .

In This Chapter

▶ Examining and describing crystals

▶ Defining crystal structures

▶ Expanding crystals into crystal systems

▶ Studying characteristics of crystals

▶ Recognizing how crystal defects are useful

. . . .

*W*hen many atoms come together, they form the large structures you use in everyday life. For example, when silicon and oxygen combine in large quantities, you get sand that you find on beaches, or glass for windows and kitchenware, or you can even make electronic and computer parts. Each example is a solid, but the applications are different, due to the differences in crystal structures between the solids which give them very different properties.

In this chapter, we start by introducing the most basic structural unit of a solid material — the unit cell. From there, we explain how crystalline solids form different shapes and are organized into different systems. At the end of the chapter, we explain some of the important and useful characteristics of solid materials, including their use as superconductors, and semiconductors.

Studying Solid Structures

Solid structures are categorized as either *noncrystalline* (also called *amorphous*) or *crystalline*. The difference between the two is how the atoms are arranged. Noncrystalline, or amorphous, solids are made up of small particles that are about 100 Å in size. The atoms composing these particles are not neatly arranged in any specific order. They lack what scientists call *long-range order,* or a repeated pattern of organization at the atomic level.

To build a crystal, you start at the smallest level with atoms or ions. For example, consider the ionic compound sodium chloride, NaCl, found in your salt shaker at home. The positively charged sodium ions are attracted to the negatively charged chloride ions (opposites attract), and they stack together into a very neat pattern. When enough of these oppositely charged ions come together, they make up a unit cell — the smallest repeating unit of the extended three-dimensional (3D) structure.

Solid structures are generally categorized as being either crystalline or amorphous. Crystalline materials are a little easier to describe because there is a lot of order in crystals. The atoms all line up in a particular fashion, and they have a long-range order. The way the atoms line up is described by a repeating unit called the unit cell; however, amorphous materials are a little trickier to describe because they can also have some repeating units within them, but they don't necessarily all line up throughout the whole material. They may be ordered in places, but over all they may not have long-range order.

Diamonds make for wonderful gifts. They're exquisite examples of a crystalline material with a unit cell that repeats itself over and over until all the crystal has a similar shape to the unit cell.

Glass is a special kind of amorphous solid, and it serves as an analog for comparison for other types of materials. For example, when proteins fold into macroscale structures, they can sometimes be referred to as being in a *glassy state.*

Building crystals with unit cells

The *unit cell* is a repeating unit of a crystal. It's the smallest volume that contains all the structure and symmetry information that upon translation represent the entirety of the crystal structure. It's like a Lego building block; the shape of the Lego block, or unit cell, is determined by the atoms, molecules, or ions and their arrangement with respect to one another. When many Lego blocks are bonded together, they represent the crystal. And the crystal shape depends largely on the shapes of the Lego blocks from which it's made.

Have you ever noticed how oranges are stacked on a shelf? You get three oranges on the base, and one that fits between them on top, and so on. This is the same for atoms in a solid, because an atom is kind of like an orange; it has a certain shape and size due to the periodic distribution of charge all around it, so it stacks up with other atoms in a certain way. No matter how big they are, they always stack up the same way to create the unit cell of a crystalline solid, or crystal. The unit cell organization of atoms repeating throughout the crystal is called *long-range order*. Because the unit cell represents how all the other atoms in a crystal are organized, you can simplify things by considering only the unit cell, rather than worrying about the trillions of other atoms in the crystal.

In short, it's the simplest arrangement of spheres, that when they're repeated, represent the whole crystal structure. A crystal is made from many unit cells bonding together.

A unit cell is a three-dimensional object, so it has six faces and eight edges, just like a cube. There are four unit cell structures that are found most commonly, these are the simple structure, the body centered cubic (bcc), the face centered cubic (fcc), and the hexagonal close packed (hcp). Figure 18-1 illustrates these four unit cells.

Figure 18-1:
From left to right is the unit cell for a simple structure, the face centered cubic (fcc), the body centered cubic (bcc), and the hexagonal close packed (hcp).

The most basic unit cell, a primitive one, is similar to a square box, with each corner of the box holding one-eighth of an atom, so there is a total of one complete atom in the unit cell. The size, shape, and other characteristics of a unit cell are described as its *crystal structure*. For example, a unit cell of NaCl, or rock salt, has a different shape and chemical characteristic than a unit cell of silicon dioxide (SiO_2) or quartz. Although they are both crystals, they have different shapes and different properties.

For example in Figure 18-2 you can see again what the simple structure looks like. In the middle of the figure you can see how the eight atoms make up the unit cell. On the right side, you can also see that inside the unit cell you have the corners of the eight atoms, and each corner of the box equals one eighth of the atom. Naturally, 8 x ⅛ = 1, so you can see that in a unit cell there is one complete atom.

Figure 18-2:
The left shows the simple structure; the middle shows the expanded view of the atoms; the right shows the one-eighth of the atom in the corners of the cell.

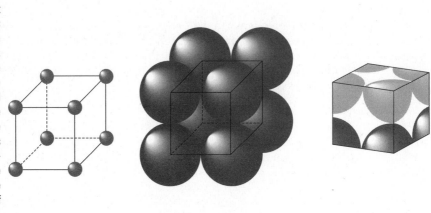

Labeling lines and corners: Miller indices

The geometric characteristics of a crystal structure are described using a tool called *Miller indices.* Because a crystal is a three-dimensional object, it must be described in three directions, similar to a 3D graph that uses x, y, and z directions to describe a point in space. In *crystallography,* or the study of crystals, instead of being called x, y, and z, the directions are called *Miller indices,* denoted by h, k, and l. This helps avoid confusion, because when you see h, k, or l, you know someone is talking specifically about crystals, and not just about 3D objects.

Any point on the unit cell can be described using h, k, and l. And the sides of a unit cell are described as a, b, and c. A plane is formed when there are multiple unit cells attached together, there are three different planes shown in in Figure 18-3.

Figure 18-3:
Three planes of a crystal: the (001) plane on the left; the (011) plane in the middle; and the (112) plane on the right.

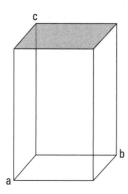
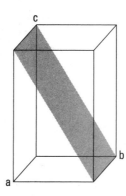
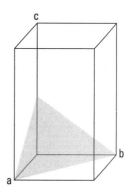

The space between the atoms is called the *d-spacing,* and it describes how much room there is between the atoms in the unit cell. The distance between points in the unit cell is denoted by the units a, b, and c. For example, if two points are being measured that are both on the h axis, then that length between them is called a. The angles between the lines are denoted as α, β, and γ. (This is shown in Chapter 8, specifically in Figure 8-4.)

Recognizing the Miller indices and crystal dimensions is important when doing x-ray diffraction (XRD). This technique measures the size and shape of the crystal, along with the positions of all the atoms in the unit cell. For more information, see Chapter 22.

Three Types of Crystal Structure

Generally speaking crystal structures fall into one of three categories: simple, binary, or complex. The number of atoms bonded together, the number of different elements involved, and the size and other characteristics determine whether the crystal formed is simple, binary, or complex.

Simple crystal structures

A *simple* crystal structure is composed of two atoms (or ions) arranged together. However, even within this simple definition there are many types of structures possible, due to the many different combinations of atoms that can fit together. The periodic table includes of a wide variety of atoms of varying size and electronegativity.

Ionic crystals are formed based on a sharing of electronic charge. So depending on what atoms are chosen, the total electron population of delocalized electrons can vary significantly, creating a variety of materials that exhibit successive changes in the electrical conductivities and melting points. Additionally, the size differences between the constituent atoms can lead to deviations of the classic unit cell to create crystals with unique macro scale shapes.

The coordination number and the geometry of the two atoms involved are the most important characteristics. The coordination number (CN) determines how many neighboring atoms can be bonded around that atom. The ratio between the CN of two atoms in a unit cell also reflect the stoichiometry; for example, Na has a CN of 6, and Cl has a CN of 6, so together the NaCl crystal has a 1:1 stoichiometry, meaning there is one Na for every one Cl atom.

Binary crystal structures

Binary compounds have two elements present. And there are many ways two atoms can fit together. Knowing the binary structure is often enough to describe the properties of the binary compounds that are talked about later in this chapter.

There are four main factors that influence what structure a binary crystal forms.

- ✔ **The size of the ionic radii:** The sharing of electrons is required to stabilize the total charge of a binary compound. However, if the atoms can't get close enough to one another, they can't share electrons effectively. The *ionic radii* is a measure of how close a cation-anion pair of ions can get to one another. To form a stable solid, the cation-anion pair needs to be as close to one another as possible.

- ✔ **The ratio of the radii:** The ratio of the radii between the cation and anion determines the structure because it affects the coordination number of the constituents. If each radii is similar, then you can expect high coordination numbers for both. The ratio is defined as r_{min}/r_{max} where r_{min} is the smaller radius, and r_{max} is the larger radius. Examples of what structures are predicted by using the radius ration rule is given in Chapter 8.

- ✔ **The polarizability of the ion:** An external electric field creates a dipole within the atom. This affects larger ions more than smaller ones. In a crystal, this happens when other ions are close by. The electric field from nearby ion's electron cloud can polarize an anion. This stabilizes the net crystal structure, because it lowers the energy of ions in the crystal. Sometimes Van der Waals' forces between ions (see Chapter 6) come into play and influence the structures that are made (this is especially the case when large ions are present).

> ✓ **Covalent bonding:** Covalent bonding occurs when there is both directionality and sharing of electrons between ions, although it's not common for ionic crystals, some amount of covalent character occurs. Other cases where this may arise include when bonding occurs between two atoms of the same element.

When there is *homonuclear* bonding, such as in the case of metal-metal bonding, there is no difference between each of the atoms. Each neighboring atom is identical, so there's consistent electron distribution about the crystal. There is no size difference, and the compound tends to form a thermodynamically stable state.

Complex crystal structures

The term *complex crystal structure* can describe crystals with multiple types of complexity. The complex structures not only have atoms on the corners of the box mentioned earlier in this chapter, but also on the face of the box, in the center of the box, or anywhere in between. Here are the most common complex crystal structures:

> ✓ **Ternary structures:** Some complex crystal structures are considered complex simply because they have three elements. In the majority of cases, ternary structures are oxides.

> ✓ **Intercalation compounds:** Alkali metals react with graphite in such a way that they get between the layers of graphite and cause the layers to expand. Intercalation compounds are commonly composed of the alkali metals, bromine, or some electron donor molecules such as amines or organometallic compounds. This is found in lithium ion batteries, for example.

> ✓ **Microporous structures:** Structures made with pores in their crystalline arrays can include different materials. As a result, they can have different applications as a result of the difference in structure and chemistry. Zeolites, for example, are alumunosilicate solids that are based on SiO_4 and AlO_4 tetrahedra framework; they contain pores and channels with varying dimension according to their desired properties. Some are used as catalysts because of the extremely large surface areas, whereas others are used as ion- exchange membranes.

> ✓ **Homoelement bonding:** Homoelement bonding is when there is bonding between two atoms of the same element. In some cases, one of the atoms can have a different oxidation state even though they are of the same element. In this case, the crystals often have a slightly anomalous stoichiometry. A special class of such crystals are known as *Zintl compounds,* and these occur when an electropositive metal bonds with a p-block element.

Calculating Crystal Formation: The Born-Haber Cycle

The bonding of polyatomic molecules can be considerably more complex than that of monoatomic molecules. For example, the ammonia molecule NH_3 has three hydrogens bonded to a central nitrogen core. Each of the three bonds must be formed in order to form a molecule of ammonia; consequently, to dismantle an ammonia molecule, one needs to remove each of those bonds. As each bond is broken, the energies of the other bonds differ. Each bond has an energy associated with breaking the bond (or forming the bond), and each one is a different value. The strength of the bonds can be affected by the environment, which can be controlled by adjusting the experimental parameters to weaken each bond so that they can be broken. Conditions such as the partial pressure of the reactant, the total pressure of the system, or the temperature of the reaction can affect the outcome. The same can be said for all polyatomic molecules, they all follow the same general rules associated with bond breaking and bond making.

The *Born-Haber cycle* was developed in 1919 by two scientists. It's based on six steps that take you from the starting materials to the crystal that's finally formed. The Born-Haber cycle is useful, because with a few simple calculations you can predict how to make crystals with different materials. The inputs needed include:

- **The lattice energy of the crystal:** This is the energy it takes to hold or break the crystal lattice apart.

- **The electron affinity:** This is the amount of energy that's gained or released when an atom gains an electron.

- **The ionization energy:** This is the amount of energy an atom needs to gain an electron from a gas or from a neighboring atom.

The First Law of Thermodynamics states that for every unit of energy that is put into a chemical reaction, the same amount of energy comes out. No energy is lost or gained in any chemical reaction. Because of this fact, the term used to calculate the Born-Haber cycle is the *enthalpy change*. The enthalpy change is a way of including all the energy in the system, including the energy in surroundings.

The Born-Haber describes the crystallization of atoms as the they undergo changes to the following:

- **ΔH_f:** Enthalpy of formation, where the elements are brought to their standard state out of the ionic lattice

- **ΔH_s:** Enthalpy of sublimation, whereby you calculate the energy required to form 1 mole of gas from the standard state

> ✔ ΔH_d: Enthalpy of dissociation, sometimes also measured as one-half enthalpy of dissociation
>
> ✔ I: Ionization energy
>
> ✔ E: Electron affinity
>
> ✔ U: Lattice energy

In sum, for sodium chloride, the crystallization can be calculated using the following equation:

$$\Delta H_f = +\Delta H_S + I + \tfrac{1}{2}H_d + E + U$$
$$381.1 = 108.4 + 495.4 + 120.9 + E - 757.3$$

And so by calculation we find that E = –348 kJ/mol.

The energy in question is called the *standard bond enthalpy*. It can be tested, and the values exist in standardized tables that can be referenced in textbooks. They can be used to determine the stability of compounds that can be either covalent or ionic. However the calculation breaks down at times, for example if values are very low positive or negative values, the results can't always be trusted. To compensate for this, a chemist can change the conditions under which the reaction takes place to better suit crystal formation.

There are six different enthalpy terms with which you need to familiarize yourself. These terms are combined in the previous master equation that sums the total of the reaction pathways, both for crystal formation and for crystal dissolution. The balance of the equation, tells you which way the reaction will go. The process is based on calculating at each step the change in enthalpy of the molecules and atoms (see Figure 18-4). When the process is exothermic (generates heat), it's given a minus (–) sign, and when the process is endothermic (requires energy), the value is given a plus (+) sign.

To use the cycle, you need to consider four properties that can be looked up in a data Table such as the ionization energy, electronic affinity, bond dissociation energies, and the heats of atomization.

To help you track yourself through each step, consider the energy required to change from a gas (g) to a solid (s). So to go the opposite way, to go from ice to steam (water vapor), you need to apply heat energy. So for all those reactions changing states from solid to gas or vice versa, you can imagine why it's endothermic (upward-pointing arrow), or exothermic (downward-pointing arrow).

The *lattice energy* is the only term that can't be experimentally determined, so it must be calculated (see Chapter 8).

Regardless of what type of crystal is formed, whether it's a covalent crystal or an ionic crystal, the likelihood of whether or not it forms can be calculated using the Born-Haber cycle.

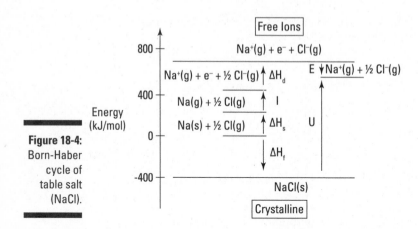

Figure 18-4:
Born-Haber
cycle of
table salt
(NaCl).

Bonding and Other Characteristics

When materials are in the solid state, they're very often quite stable because there is little wiggle room for atoms to move around. In other words, reactions between solids can be extremely slow; because of the high energy barrier put on the diffusion of atoms, they're locked into the solid state and can't move about freely. Because of this, solid state materials must often be made at high temperatures to overcome this energy barrier, but when cooled, many solid state materials can remain intact for a very long time.

When atoms are in the solid state, they can be tested using various techniques, and it's often the best way to determine certain properties of the elements because they're not moving around — they're fixed into a physical position and chemical state. For example, the oxidation states based on the coordination number and the size of the atoms can be tested this way.

When solids are formed, they are very stable, but they can be dissolved into their individual constituents by using an appropriate solvent. Just think about how a solid lump of sugar dissolves into a mug of hot coffee.

When in the solid state and the atoms are all bonded in a tight-knit manner with each other, they don't move around very much; however, they're not totally clamped into position. When solid materials are heated, the atoms get excited and start to vibrate. When they vibrate together they sometimes rub up against each other and the friction causes them to heat up.

When materials are super-cooled they respond in the opposite manner; they become less excited, tend to settle into their position, and do not vibrate very much. At the super-cool temperature of 4 Kelvin (–269 °C, –452 F) electrons can travel very fast in the solid material and experience nearly no loss of energy, this property is known as superconductivity and is known to occur for a variety of elements. It's covered in greater detail in this chapter in the section "Encountering zero resistance: Superconductivity."

There's also another situation that arises in solid state materials. Remember that in crystals many atoms are lined together, the electrons experience a collective level of energy known as a band. They can reside in either the valence band or the conduction band, depending on the chemistry and state of the material. The energy difference required to raise an electron from the valance band to the conduction band is called the bandgap energy. This is an important concept in solid state chemistry; it's used to classify semiconductor materials and photovoltaic materials.

Characterizing size

The most common method used to characterize crystals is X-ray diffraction. This is carried out by shining a strong beam of X-rays at a crystal and then detecting how the pathway of the beam changes (called *diffraction*). The reason that X-rays are used is because their wavelength is the same size as a typical chemical bond, so it's sensitive to the differences in chemical bonding within a crystal.

X-ray diffraction is described by *Bragg's Law of Diffraction*. The equation is pretty simple:

$$n\lambda = 2d \, \mathrm{Sin}\theta$$

where λ is the wavelength of the x-ray ($Cu_{k\alpha}$ = 1.54 pm), d is the distance between the crystal planes in the crystal lattice, and θ is the angle that the x-ray beam gets sent off course, or diffracted.

As you can see on the left in Figure 18-5, combined waves interact constructively at the particular 2θ angle and will, therefore, have a high intensity at the detector. But on the right, the wave undergoes a destructive interaction that cancels out the wave, which means there is no x-rays hitting the detector.

It's the electrons in the atoms, that cause the x-rays to diffract. Because it's the electrons, the diffraction can also tell you something about the coordination of the atoms in the solid, and this in turn can show what class of crystal it is. The CN can help distinguish between ionic and covalent solids. For example, ionic solids often have low CN.

Figure 18-5:
On the left, the x-rays pass through a crystal and are deflected coherently into a detector; the image on the right shows x-rays passing through a crystal at a different angle and they cancel each other out.

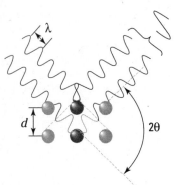
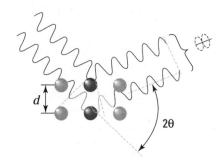

You can determine the structure of your crystal by noting where the peaks are located. If you have a very large crystal with lots of order, the peaks are very sharp. But if you have small crystals, like for example in the case of nanoparticles, the peaks are broadened. For more information about nanoparticles see Chapter 19.

In other cases you can also use electrons instead of x-rays to shine through materials. This works only in a vacuum because electrons are easily scattered in air, and if the material is so thin that the electrons pass through to the other side. This can be done using a transmission electron microscope or a particle accelerator.

The principle is the same, but you get a black and white image. It's based on the diffraction of light waves when passing through a periodic medium that acts like a diffraction grating that influences the coherence of the electron beam.

Dissolving in liquids: Solubility

The term *solubility* describes the property of a *solute* (usually a solid material such as a metal salt or some kind of organometallic compound) to dissolve into a *solvent* (usually a liquid), including how much of the solute can remain in the solvent. When you add table salt to water, the table salt is the solute, and the water is the solvent. How well the solvent can take the solute, is called *solvation*.

There are two type of solvation interactions: a *specific solvation*, and a *non-specific solvation*. In both cases there is some interaction between the solvent and the solute.

In the case of specific solvation, there's a direct interaction between the solvent and one or more of the ions within the solute. This can be in the form of binding between the solvent molecules with the solute, for example in the case of the inorganic salt copper (II) ammonia complex, $[Cu(NH_3)_4]^{2+}$.

In nonspecific solvation there's a weaker interaction that arises due to either van der Waals forces (see Chapter 2), or dipole-dipole interactions between the solute and the solvent. This is what happens when you dissolve salt into water.

There are four types of solvents:

- ✔ **Polar solvents:** A polar solvent is polar because of an uneven distribution of charge about the molecule, meaning one side is more positively charged than the other side. The most typical polar solvent is almost certainly water. Something more like ethanol is used because it's slightly less polar.

- ✔ **Nonpolar solvents:** Nonpolar solvents are usually hydrocarbon molecules that have an even distribution of charge on the molecule. These are commonly used in organic chemistry and organometallic chemistry because many organic molecules are not polar. Examples include benzene and hexanes.

- ✔ **Protic solvents:** They can donate protons (H+) to the solute, and this can be useful because it can change the charge on the solute by the addition of a proton or positive charge. This can help stabilize the solute and make it more soluble.

- ✔ **Aprotic solvents:** This doesn't donate protons and can be used when there's no need for protons to be added to the solute.

In general, solvation is governed by the ability for donor-acceptor interactions to occur between the solvent and the solute. Strong polar solvents are good at solvating charged species. Good solvents have donor (Lewis base) and acceptor (Lewis acid) properties, because there's a form of bonding that occurs between solute and solvent. More details on Lewis acids and bases can be found in Chapter 5.

Solvation of ionic materials is based on an equilibrium constant known as the *solubility product*. It arises from thermodynamic considerations and is directly related to the Gibbs free energy change (see Chapter 16). This in turn is related to three components: lattice energy (see Chapter 8), solvation energy (see Chapter 8), and entropy (see Chapter 9).

Encountering zero resistance: Superconductivity

When many metals and alloys are suddenly cooled to liquid helium temperatures they have *zero resistivity;* in other words, they act like a *superconductor.*

Superconductors transmit electricity and effect magnetic lines. Conversely, they can also be destroyed by magnetic fields. They were discovered in 1933 when a metal material was cooled below the critical temperature, T_c, when inside a magnetic field, the metal expels the magnetic fields, causing it to levitate above the substrate. Some people have even considered using this as a way to operate train tracks so they can travel great speeds. However, if there is a really strong magnetic field present, it can lower the T_c of the material, helping it to become superconductive.

The nature and origin of superconductivity was described in 1957 by John Bardeen, Leon Neil Cooper, and John Robert Schrieffer. Together they created the Bardeen Cooper Schrieffer (BCS) model. It occurs for many metals, alloys, intermetallic compounds, and doped semiconductors. The transition temperatures range from 92.5 K for $Yba_2Cu_3O_{6.9}$, down to 0.001 K for the element Rh. And there are some materials that become superconducting only under high-pressure conditions. These materials all have to be extremely pure, even just one impurity in 10,000 atoms can severely affect the superconducting property.

The mechanism of superconductivity is still under investigation, but BCS theory is the currently accepted model. In short, what the model says is that the crystals line up very well in the pure material, with little to no defects present. The oxidation of copper is somewhere between +2 and +3, and that free electrons join up in the crystal and form what is known as *Cooper pairs.* As a pair of electrons they are less likely to scatter when they come across a defect in the crystal, and so they can travel very far and very fast. This allows the current from an outside potential to be transported with high efficiency.

The BCS theory doesn't fully predict what other materials will act as superconductors, and so there is room for improvement through further scientific research. Ultimately, scientists hope to have a material that can be used at room temperature to transmit electricity over great distances. Until that's developed we have to settle for another type of conductor that works on *ballistic conduction* instead of superconduction (see Chapter 8 read up on ballistic conduction).

Information technology: Semiconductors

In metals, conductivity is permitted because the delocalized electrons are all residing in the conduction band, meaning they're available for electrical conduction. But in nonmetals, or semimetals such as silicon, the electrons are in the valence band, and there's a bandgap that must be overcome for the electrons to get into the conduction band. The materials can conduct when that energy is supplied, which is why they're called semiconductors — because they can conduct only under the right circumstances.

In short, the *band theory* looks at the three-dimensional object as being composed of many metal atoms that each have overlapping molecular orbitals. The sum of all the overlapping orbitals gives the metal a collective band where the electrons reside, and the band level determines the energy of the electrons in the band.

Freezing Fermi

Enrico Fermi was an important Italian scientist in the early 1900s who thought long and hard about what would happen if you were to freeze an atom to absolute zero. He asked questions such as:

✔ Would it stop vibrating?

✔ If it stopped vibrating, would that mean it had no energy?

✔ If it had no energy, would that mean based on the equation $E = mc^2$, would it have no mass?

✔ If it lost all the mass, would that mean it would just disappear?

So he calculated what would happen and found that the atom would continue to vibrate even at absolute zero. It would have some small amount of energy left, so it would not just disappear. This energy of the atom, still present at absolute zero, is called the *Fermi energy*. Be careful not to confuse the Fermi energy with the Fermi level. Although they're same at absolute zero, they're not the same at elevated temperatures.

Because the Fermi energy is the energy at absolute zero, the Fermi level represents what that energy would be under certain conditions. It's often more practical for this reason and is used frequently. The Fermi level gives a practical understanding of how much energy is required to fill the gap in the electron band; in other words, it can be used to describe the bandgap.

In the band structure there's a *valence band* (VB), and a *conduction band* (CB). In metals, the VB and CB overlap, but in semiconductors there's some energy gap between the two and the size of this bandgap determines how much energy is needed to promote an electron from the VB to the CB.

Metallic solids have a continuous band of electronic energy levels with the Fermi level being filled (see the following "Freezing Fermi" sidebar), but non-metallic solids have a bandgap at the Fermi level. Bandgaps mean that the solid have some electrical resistance, because there's extra energy required for an electron to go from the valence band into the conduction band. Size of the gap determines how much resistance it has. Depending on the elements used, the extra energy required to go from the valence band into the conduction band can come from light. The bandgap determines optical properties of materials — those properties are used to trigger light switches for that reason, and they have electrical resistance so they are used in semiconductors to create transistors for example.

Synthesizing Solid Structures

Inorganic chemists use a variety of tools and methods in the preparation of their materials. They use gas-phase chemistry where materials are vaporized and then condensed into crystals. Or, they use high-energy sources like microwaves to supply the energy needed to bond atoms together. But, for the most part, chemists use liquid phase reactions because the temperatures are more amenable, and there is a high degree of control that can be afforded. Here are just a few of the ways different solid structures are made in laboratory settings.

- **Homoelement materials** can be made a number of ways. In the semiconductor industry, silicon chips are made by evaporating silicon compounds down onto a crystal template that helps seed the crystal growth.

- **Ternary complexes** can be made in the liquid phase where the compounds are reacted at the appropriate temperatures and ratios.

- **Intercalation** can be carried out in the liquid or in the gas phase. The intercalation atoms need to have some wiggle room to move between layers; after they get between the layers of the material, they cause it to swell.

- **Microporous structures** are often made in the catalysis industry because they can be designed with specific pore sizes. They can be readily made by mixing the appropriate compounds in a liquid solvent, then upon drying they are often annealed at a high temperature that has the effect to seal the bonds between the atoms.

✔ **Nanoparticles** are crystals of metals that have physical dimensions at the nanometer scale. They're interesting because they sit at a size regime that's governed by a crossover of quantum mechanics and classical mechanics. Some surprising and useful properties have been discovered only just recently. As a crystal, the nanoparticles could grow larger and larger but instead they are kept at a certain size because of the addition of ligands around the surface. The ligands compete with the metal and prevent metal from bonding to the outside. In this way nanoparticles of various sizes and shapes can be made. It's similar to the kinds of things that happens in organometallic chemistry. (For more information on the bonding of metals and ligands, look at Chapter 15; see Chapter 19 for more about nanotechnology.)

Detecting Crystal Defects

When talking about crystals, the word *defect* means that the regular long-range crystal structure is altered. Although this can seem bad, it's sometimes a useful property, especially in the semiconductor industry.

Defects come in many shapes and sizes. When the regularity of the periodic lattice is broken, it can be in the form of a line defect, a plane defect, or a point defect.

✔ **Line defect:** A line defect can occur when one crystal plane doesn't extend uninterrupted all the way through the crystal. It's like having a stack of paper with one sheet slightly shorter than all the others. The paper above and below that shorter sheet experiences a slight strain because of the change.

✔ **Point defect:** Point defects are important for the semiconductor industry and are classified according to whether they have vacancies, interstitials, or impurities. They're used to dope semiconductors for use in information technology industry.

✔ **Plane defect:** Plane defect forms a plane that is cleavable because both planes don't match up at the defect location. Imagine a deck of cards where half the deck are sideways. The plane at which the two halves meet is the plane defect.

Chapter 19

Nanotechnology

*N*anotechnology deals with the science of very small things that can make a very large impact on society. It's a modern field of study, but has been around for quite some time. Nanotechnology incorporates all aspects of chemistry so that materials with nanoscale dimensions can be made, assembled, and understood. Nanomaterials have many unique and fascinating properties, and these are often very different from the properties of the same materials when in they are in the macroscopic phase. Scientists have only begun to scratch the surface of the science of nanotechnology, and there are many more exciting tools, remedies, machines, devices, and scientific understanding yet to come.

Defining nanotechnology

The word *nano* comes from the ancient Greek word that means *dwarf,* so as you can imagine nanotechnology deals with the science of very small objects. To be precise, nanoscale ranges from 1 – 100 nanometer (nm), where one nanometer is 1×10^{-9} m; this is one billionth of a meter. Nanotechnology is slightly larger than atoms (an atom's size is measured in angstrom (Å) where one angstroms is $1/10^{th}$ of a nm). So a nanometer is about ten times bigger than an atom.

On the larger side of the size scale, at 100 nm, which is considered the upper limit for nanotechnology, you start to get into the realm of microtechnology, which is used in computer chips, for example. A micrometer, also known as a micron, is 1×10^{-6} m, so 100 nm nanomaterials equals 1×10^{-7} m.

So in effect, nanotechnology is ten times bigger than atoms and has a range of sizes that spans from the size of DNA molecules, all the way up to sizes that are used for computer chips. For this reason it's often best understood by knowing about all of chemistry, physics, and biology. And for this reason, it can be important in the understanding of all of these fields of study, too.

History of nanotechnology

The word *nanotechnology* has only been recently introduced. It's a relatively modern way to study materials, even though people have been aware for a very long time that materials with these kinds of dimensions (1 to 100 nm) have some very interesting properties.

In this section, we focus on four points of interest because they lead the way to all that nanotechnology can achieve at this current time.

Imagine this: It's Christmas 1959. A famous and exuberant physicist named Richard P. Feynman gave a lecture entitled "There is plenty of room at the bottom." In this talk, he set out a scenario in which atoms could be manipulated one by one, they could be put in places they were needed, and molecules could be made that would act like parts of machines. He suggested it could be possible to make a machine, that in turn could make a smaller machine, that in turn could make a smaller machine, again and again until you eventually have a machine that was just moving atoms around. For many people, this was — and still is to this day — a great inspiration. He realized and highlighted publicly that there were no physical reasons, no reasons that he could see (and he was very smart . . . a Nobel laureate, no less) why his machine couldn't be made. This gave inspiration to inventors and scientists all over the world to try to make the machine he was talking about.

Then in 1974, the word *nanotechnology* was coined by a Japanese professor named Norio Taniguchi when he talked about the specifics of making materials for the semiconductor industry. With a name, it was much easier to refer to it as a single line of study, and so people started to focus on what can be done at the nanoscale.

Then in 1989, scientists were able to pick up and move one atom at a time using a device called a scanning tunneling microscope. They were able to neatly arrange xenon atoms and spell out the word "IBM." Now it seems that nanotechnology really could be possible, and everything that Feynman talked about could become a reality. Two scientists, Gerd Binnig and Heinrich Rohrer, received the Nobel Prize for developing the scanning tunneling microscope in 1986.

An interesting molecule was later discovered by three researchers trying to make carbon molecules that they believed were found in space. They discovered fullerenes in 1985, a soccer-ball-shaped shell structure forming a molecule comprised of 60 carbon atoms, so it's referred to as C_{60}. It has the structure of a geodesic dome, like an icosahedron proposed by Buckminster Fuller, thus also commonly known as the Bucky Ball and the fullerene family of molecules. For this discovery of a new allotrope of carbon Harold Kroto, Robert Curl, and Richard Smalley share the Nobel Prize in 1996 for Chemistry. This molecule was interesting because people started to imagine that the inside cavity could be used to carry cargo, like drugs, that can then be used for targeted delivery within the body. Later on, a new material, that is part of the Fullerene family was discovered that was called a *carbon nanotube*. A daughter molecule to fullerenes, it boasts incredible strength and remarkable properties and a whole slew of new materials were envisioned.

Carbon has been used since the beginning of time; however, the discovery of a new elemental allotrope sparked great interest. In short, because even a material like carbon, that we thought we knew everything about, can have such radically different and exciting properties when it is constrained to the nanoscale. This begs the question: What other nanomaterials can be made that suddenly have such new and exciting properties? This is what inspires many people who work in the field of nanotechnology.

The science of nanotechnology

At the size scale that we live in — the big stuff that we see in everyday life (the macro world) — there's a set of physics laws that we use to help us understand the processes around us. These are called the *laws of classical physics,* that allow us to understand tides, the effect of gravity, and lots more. You can test these theories in your backyard, if you want to.

But at the atomic scale, there are a different set of physics laws at play. The rules at that size scale are governed by quantum mechanics, and these help us in chemistry to know where an electron is most likely to be found around an atom and what the orbitals look like. But a lot of other funky things happen, too; things like quantum mechanical tunneling, where an electron can traverse an potential barrier and suddenly appear somewhere it is not supposed to be able to be reach.

Top-down versus bottom-up

From the first tools that were made, by carving pieces of fish bones or chiseling rocks to form axe heads to modern day computers, humans have used the same top-down design philosophy. For example, in making a computer chip, an ingot of pure silicon is etched with light, pulverizing the substrate and blasting away channels that become conduits for electricity and electrons. But it's becoming more and more clear that we can make materials in a way that is similar to how nature works. Chemists can make small chunks of some device and put them together, and under the appropriate conditions they can self assemble. For example, small groups of amino acids can be joined together by covalent chemistry, and these can act as building blocks. They are then placed in an appropriate medium, and under the right conditions the building clocks can self-assemble into fibers. The fibers can then be used for a whole slew of applications.

The essential difference between the two approaches is that with bottom-up design, you design the smaller parts — the building blocks. You design them in such a way that under the right conditions they assemble together and make a much bigger structure. Just like a house is made of bricks, scientist make the bricks, and then find the right conditions so all the bricks align and form a house.

Nanomaterials

By definition a nanomaterial has at least one component having a size scale between 1 and 100 nm. More often than not, when a material is at this size scale it tends to have quite unique and special properties that make it different from when it's in the bulk or macroscale. It's by making materials at the nanoscale that new and exciting properties are discovered. After those properties are discovered, special applications can be found for them and they can be used for a variety of uses.

Nanomaterials have some interesting and useful properties. They have extremely high aspect ratios, meaning that the surface area compared to the volume is very large. This is useful because the surface is where chemistry happens, so the more surface area that you have, the more reactivity you can achieve.

They're also colloidal. They can be easily dispersed in a liquid solvent, so they can be used in liquids, or they can be used as an aerosol. Basically, they're just the right size to be easily distributed either in liquids or in gasses.

Size and shape control

Depending on the metals used, nanoparticles have certain crystal lattices (see Chapter 8 and Chapter 18) that can determine what the overall structure of the material looks like. Large-scale crystals can have lattices that extend for a very long distance, but in nanoparticles the crystals only extend up to 100 nm. To stop the crystals from getting larger and larger, forming microparticles instead of nanoparticles, you can use ligands to bind the face of the crystal. (For examples on how ligands work see Chapters 9 and 15.) When the ligand binds to the face of a crystal, it competes with the addition of more metal atoms. This can stop more metal atoms from binding to the crystal, and in this way both size and shape control of nanoparticles can be achieved.

In making nanoparticles, it's common to find that a metal precursor such as silver nitrate is used. The silver and nitrate aren't covalently bound, and so when they are added to a polar solvent like water they each dissolve readily. The silver can then be reduced using a simple reducing agent such as sodium borohydride, and when this happens the silver atoms start to bind together and crystallize. If an appropriate ligand is also employed in the reaction — for example, tri-sodium citrate — the nanoparticles can be made into specific shapes. The citrate is loosely bound to the surface of the nanoparticles and can be used to shape them. The list of nanoparticles shapes is extensive. The most basic and common shape is spherical because the surface energy is equal around the surface. But it's possible to make cubes; when you have a cube you increase the surface energy because you have to hold atoms in place and stop them from spreading out and allowing the nanoparticles to morph into a sphere. Other shapes include rods, stars, triangles, tetra pods and more.

With each new shape, the nanomaterials can have different and unique chemical and physical properties that are different from the bulk material. Scientists have only just begun to make nanoparticles from a small handful of chemical elements. You can expect many new technologies to be borne out of making interesting nanoparticle shapes with other elements from the periodic table.

Self-assembly and gray goo

Nature uses self-assembly all the time, yet scientists and engineers are light years behind what nature can do. Self-assembly deals with the science of making objects that put themselves together, just like a seed can grow into a massive tree by converting all the materials around it into the wood that becomes the tree.

DNA origami has been made by designing the correct sequence of amino acids together to form single DNA strands, that bind together in elaborate and unnatural ways — unnatural in the sense that so far this has not been found in nature. You see, DNA is a helical structure that's bound in the center using base pairs that are like the rungs of a ladder. There are two possible pairs allowed, and they're made from either guanine (G) and cytosine (C), or thymine (T) and adenine (A). Just think of the base pairs like the 1s and 0s that make up computer that's used in the information technology industry today. It's this complementarity in DNA that makes it have the capacity to pass on information, the kind of information that makes you *you*. But when the DNA sequences are altered slightly, they can be made to fold in rather different ways, and so DNA origami can be created.

Nanotechnology has been trivialized in the media, somewhat thanks to some entertaining science fiction. One idea to emerge suggested that nanotechnology could create some kind of gray goo that's composed of trillions of small nanobots or nanomachines. The idea is that these bots or machines could feed from their surroundings and self replicate until they turn everything into a gray goo. It helps to get people thinking about what is possible by investigating nanotechnology, but as of yet the idea is not in line with what is known in the scientific field. However, one of the greatest leaps made in this direction involves a company that was able to synthesize DNA that was able to self replicate. What's amazing is that the DNA sequence was picked by a computer. In essence, the traits of life were selected according to a computer algorithm that carefully chose the right sequence of amino acids. So in a sense the DNA was made using a computer program, put together using pretty simple chemistry, and then was able to self replicate. But the requirements for such a process to take place are condition-dependent, pH and temperature have to be closely controlled; otherwise, the system doesn't work. If there is a gray goo out there, so far this is the closest we know about it.

Applications for Nanotechnology

In the words of Dr. Wade Adams, "Nanotechnology is making small stuff do big things (and then selling them!)." In this way nanotechnology can make a positive impact in the world, because neither science nor technology can do good unless it's available to the world. Here's a short list of how nanotechnology can better impact the world.

Cancer therapy

Finding a cure for cancer is a very ambitious and worthy goal, to date there doesn't seem to be a straightforward method that can be used to fight off cancer. The current treatments are often so poisonous to the person who is taking them that they are just as badly affected by the medicines as they are by the cancer itself.

There are many reasons for the formation of cancer, and there are many different types of cancers, each with unique and specific influences on the body.

But they all share at least one property in common, compared with normal healthy cells in the body — cancerous cells are quite leaky and not very stable. They're leaky because they grow so fast and don't have strong cell walls. Modern treatments using nanotechnology try to exploit this common feature of cancerous cells so that they can broken down and then removed from the body.

Several candidates for this have been suggested, including using single-walled carbon nanotubes, metal nanoparticles of gold, silver, and iron, or by using other nanoparticle structures such as gold nanoshells, for example. These materials often react very strongly in an electromagnetic field that can be composed of light or radio frequency radiation, so much that they can heat up rather quickly. The idea is that the nanoparticles can be tethered to biomolecules that bind only to the cancer cells and nowhere else. When they are all in place, the patient is then treated with either a laser light or some form of electromagnetic radiation. The cancer cells get cooked by the rapidly heated nanoparticles, and they break apart and disintegrate. Eventually the body excretes the broken up pieces through the usual routes. The healthy cells are not badly damaged because they are much stronger, and also because the nanoparticles are attached only where there is cancer; healthy cells are generally left alone and don't get cooked. This would make the treatment far easier for the patient to bare.

Catalysis

Catalysis is a surface science. In heterogeneous catalysis (see Chapter 16) we see that catalysis happens on the surface of some kind of a substrate. Surfaces are important, and we also see how the total surface area is also important; when there is more surface area available, typically more catalysis occurs. For this reason nanoparticles are ideal candidates for catalysis, because they have a very high aspect ratio and a lot of surface area. In some cases the combined surface area of a single spoonful of nanoparticles can equal the surface area of a football field.

Smaller nanoparticles have very high aspect ratios; the aspect ratio is a measure of how the surface area compares to the volume of the material. Because catalysis is a surface science, ultimately more effective surface area is important for this industry.

The enhanced surface energy can lead to interesting chemical reactions, and it's even more intriguing when they occur for metals that are usually inert. Take gold as an example; gold jewelry can maintain its shine for a very long time because it doesn't react with the air around it, so it doesn't oxidize. For a long time gold was considered to be rather boring for catalytic reactions because it was so unreactive. But when the gold is made into nanoparticles, because of the added surface energy, and the strain on the surface that makes it more reactive, it can now become quite useful as a catalyst. The surface strain adds energy to the system that in turn helps catalyze chemical reactions.

Education

Nanotechnology is a cool and exciting new field of study that combines knowledge from all of the experimental sciences. This book is about inorganic chemistry, and there are many laws and trends that are talked about in this book that are useful for nanochemistry. But there's some fascinating physics happening in nanotechnology, too, like cloaking devices that use single-walled carbon nanotubes. They are able to bend light and so it can hide objects behind them. They were put together by chemists, but you need a good grasp of physics to understand how they work.

There are also interesting and exciting examples from the biological sciences; scientists and engineers are making materials that mimic nature in many ways. They are developing materials that hold together and help skin formation. This can help with healing processes and combat the onset of disease or illness by preventing infections from occurring on open wounds, for example.

All of the examples and new technologies arising because of nanotechnology are exciting, and they are at the forefront of science. In nanotechnology you find examples where science meets science fiction, and the only limits to this field are bounded by the imagination of those involved. For this reason, alone it's a great tool for education, because it can engage students to think about possibilities that previously might have been considered impossible. For example, many people may have heard about a space elevator that can take people to space along a very lengthy and strong cable. Just recently a breakthrough was made in producing extremely strong and lightweight fibers that could be used for such a cable. These are made of single-walled carbon nanotubes; the fibers are as strong as steel, but as flexible and malleable as cotton. Just this time last year such an object might have easily been considered a pipe dream, so just think of what other pipe dreams are going to become a reality soon. But they won't become a reality if students don't care about science. Hopefully nanotechnology can act as a catalyst to get children dreaming and working on making more pipe dreams come true.

Part V
The Part of Tens

The 5th Wave   By Rich Tennant

"Okay—now that the paramedic is here with the defibrillator and smelling salts, prepare to learn about covalent bonds."

In this part . . .

This part takes a look at some of the organic-chemistry-related common household products you come across in your everyday lives. Also, you find information about ten great discoveries that have been awarded the Nobel Prize. And you can find out about the instruments and techniques that an inorganic chemist may use. Then if you're feeling adventurous, you can try out a few simple chemistry experiments.

Chapter 20

Ten Nobels

*T*he Nobel Prize Foundation was created in 1895 at the behest of Alfred Nobel in his last will and testament. Nobel was a chemist; he invented dynamite and earned his fortune by this. Appalled by its use on the battlefield, he had his great wealth turned into the Nobel Prize Foundation.

The Nobel Prize has been awarded each year since 1901 to recognize great achievements in chemistry, physics, physiology, medicine, literature, or the pursuit of peace. Everything talked about in this book comes as a result of hard work and research by chemists who have (hopefully) been honored and recognized for their contribution. Some have made such a great contribution to science in their own lifetime that they're awarded a Nobel Prize. In many cases, the contribution of a hard-working scientist is not recognized until late in their career (if ever at all).

In this chapter, we describe ten Nobel Prizes for which chemists labored long hours in a laboratory. These scientists have taken chemistry to the next level and the breakthroughs in their fields earned them a Nobel Prize.

Locating Ligands: Alfred Werner

In 1913, the Nobel Prize was presented to an inorganic chemist for the first time ever. Swiss chemist Alfred Werner won the Nobel Prize for recognizing that the central metal atom of a complex is surrounded by connecting molecules (or ions), now called *ligands*. This led to the new field of organometallic chemistry and laid the foundation for further understanding of coordination chemistry (see Chapter 15). The official reason for his win was "in recognition of his work on the linkage of atoms in molecules by which he has thrown new light on earlier investigations and opened up new fields of research, especially in inorganic chemistry."

Making Ammonia: Fritz Haber

In 1918, Fritz Haber won the Nobel Prize "for the synthesis of ammonia from its elements." In other words, Fritz Haber was the first chemist to create ammonia in a laboratory. His discovery made it possible to manufacture industrial-sized quantities of ammonia that could be added to fertilizers. (Check out Chapter 17 for details on why ammonia is a useful fertilizer.) Manufacturing ammonia led to increased crop yields and transformed farming, shaping the production of foods you eat today. It's arguably the most important development in agriculture since the development of irrigation (6,000 BC) and the use of draft animals (4,000 BC).

Creating Transuranium Elements: McMillan and Seaborg

In 1951, two American scientists shared the Prize "for their discoveries in the chemistry of the transuranium elements." They literally created atoms of new elements, and then worked to understand the chemical properties of these elements. The elements they created in the lab are elements with atomic numbers greater than 92 (the atomic number of uranium) and are unstable — they radioactively decay to different elements. Very few of these elements are present in nature, but McMillan and Seaborg were able to create them in a laboratory. See Chapter 4 and 14 for more information.

Adding Electronegativity: Pauling

Linus Pauling is the only person to have been awarded two undivided Nobel Prizes. His first prize was for chemistry in 1954 "for his research into the nature of the chemical bond and its application to the elucidation of the structure of complex substances." He developed the idea of the *Pauling Electronegativity,* which helps quantify chemical bonding between atoms. He was later honored for his work regarding arms control and disarmament; he won the Nobel Peace Prize in 1962.

Preparing Plastics: Ziegler and Natta

In 1963, Karl Zeigler and Guilo Natta earned the Nobel Prize in chemistry for their work with polymer, or plastic molecules. Their discoveries in the laboratory paved the way to the modern world as you know it — a world filled with plastic! Their work set the stage for making modern plastic materials and they won "for their discoveries in the field of the chemistry and technology of high polymers."

Sandwiching Compounds: Fischer and Wilkinson

Two chemists, working independently, were awarded the Nobel Prize in chemistry in 1973. Both Ernst Otto Fischer and Geoffrey Wilkinson shared the 1973 Prize "for their pioneering work, performed independently, on the chemistry of the organometallic, so-called sandwich compounds." The discovery of sandwich compounds (see Chapter 15) ignited the imagination of chemists working with organometallic compounds. Since then, the textbook (coauthored) by Wilkinson on advanced inorganic chemistry has become standard reading for all inorganic chemists; so much so, that Wilkinson refers to it as "The Bible."

Illuminating Boron Bonds: Lipscomb

As we explain in Chapters 12 and 15, boron is an interesting atom that undergoes fascinating bonding arrangements. In 1976, American chemist William Lipscomb was awarded the Prize "for his studies on the structure of boranes illuminating problems of chemical bonding." Lipscomb was a student of Pauling (mentioned earlier in this chapter) and published a book as well as many articles on the bonding of boron hydrides. Check out Chapter 12 for details on these hydrides.

Characterizing Crystal Structures: Hauptman and Karle

The most recent Nobel Prize awarded to inorganic chemists (1985) was given to American chemists Herbert Hauptman and Jerome Karle "for their outstanding achievements in the development of direct methods for the determination of crystal structures." Together, they married chemistry and mathematics so that crystal structures could be easily identified and characterized. The work they did was carried out in the 1950s and was not accepted or truly recognized until much later. Today, many of their protocols are used on a regular basis.

Creating Cryptands: Jean-Marie Lehn

The work of Jean-Marie Lehn has led to a new subfield of chemistry called supramolecular chemistry. He studied the binding between metal atoms and organic molecules and eventually created cryptands to bind with metal centers (see Chapters 9 and 15 for more details). In 1987 he shared the Nobel Prize with Donald J. Cram and Charles J. Pedersen "for their development and use of molecules with structure-specific interactions of high selectivity."

Making Buckyballs

In 1996 Richard Smalley, Robert Curl, and Harold Kroto shared the Nobel Prize for their discovery of the buckminsterfullerene. It's a spherical molecule resembling a soccer ball, and it's composed of 60 carbon atoms. (It's often referred to as C_{60} for this reason.) The discovery led toward the science of nanotechnology. Today, Robert Curl can be found cycling to work on his bicycle. It's a sight to behold because here's a man awarded a Nobel Prize for discovering a carbon molecule, yet he keeps his carbon footprint at an absolute minimum.

Chapter 21

Tools of the Trade: Ten Instrumental Techniques

· ·

In This Chapter

▶ Using light waves, x-rays, and lasers to probe materials

▶ Providing images and clues to determine structure and composition

· ·

The practice of any skill or art requires some basic knowledge of the tools at hand. For inorganic chemists, knowledge about the elements of the periodic table form the basic toolset; to know the elements and how they react is to know how to use them to your advantage. Thanks to the development of the instruments mentioned in this chapter, a great deal has been learned about the nature of the atoms themselves and the various ways in which they interact with other molecules through various bonding interactions.

Absorbing and Transmitting Light Waves: UV-vis and IR

Both ultraviolet-visible spectroscopy (UV-vis) and infrared (IR) are spectroscopy tools that cover a large range of the electromagnetic spectrum. A light source is shined on a sample and, depending on the property of the material, it may absorb or transmit the energy at specific wavelengths. The wavelengths that can be tested span from the UV to the IR. UV light has short wavelengths and is very energetic, and the IR waves are longer wavelength and are less energetic. IR is used extensively to monitor chemical reactions because the chemical bonds vibrate at similar frequency to IR. In absorption spectroscopy atoms produce line spectra, whereas molecules produce band spectra.

Catching Diffracted Light: XRD

X-ray diffraction (XRD) helps scientists determine the structure of crystalline materials and the stoichiometry of the materials within the crystals. It's based on Bragg's Law of Diffraction ($n\lambda = 2d\ Sin\theta$) that describes how a crystalline material diffracts light waves when they pass through it. Crystalline materials cause diffraction because of the regular and periodic placement of atoms in the crystal. The crystalline and ordered structure breaks up the solid stream of x-ray waves into fragmented streams of x-ray waves according to the crystal order in the material.

Many pure crystals have been tested by scientists so a library of data now exists that can be used as a comparison against the values obtained by the experimenter.

Rearranging Excited Atoms: XRF

X-ray fluorescence (XRF) determines the elemental composition of materials and gives understanding as to the stoichiometry of the elements that are present. This is done by shining a high-energy x-ray beam onto the sample, which excites the electrons and causes some rearrangement to occur. Some electrons get ripped out of or ejected from their electron shell. The ejected electrons can be picked up by a detector. Ejecting the electron causes the material to undergo *fluorescence,* and the energy or wavelength of the fluorescence can be used to determine what atoms are present, and the intensity can indicate the concentration of the atoms present. To work properly, the instrument must be in a vacuum environment. Electrons don't travel far in air and won't reach the detector unless they are in a vacuum.

Measuring Atoms in Solution: ICP/AA

Inductively coupled plasma atomic absorption (ICP/AA) spectroscopy is used to determine the concentration of analytes in solution. It can measure concentrations as low as a few parts per billion. This method is commonly used to measure the atomic concentrations of mineralogical samples. The analytes must be in solution to be sprayed into a chamber and excited by a plasma. Upon relaxation energy is released with a specific wavelength. The wavelength corresponds to what atoms are present. The intensity of the energy is related to the concentration of the analyte present. ICP is quantitative, but it must be referenced against a set of standards that are used for comparison.

Detecting Secondary Electrons: SEM

Scanning electron microscopy (SEM) is used to visualize materials and give a picture of what they look like. Samples are irradiated with a primary beam of electrons that raster scans the material; this excites the atoms in the material so they emit secondary electrons and backscattered electrons.

The secondary electrons are collected by a detector and processed to create a picture of the material similar to what you might see looking through a microscope. But when the back-scattered electrons are detected, you can tell the difference between types of materials because back-scattered intensity is proportional to the Z number of the atom. With both pictures present it's possible to give an answer as to how the surface looks and also what way the material is arranged, based on different contrasts using the back scattered electrons. Typically, samples need to be conductive, or at least be a semiconductor, to work using SEM. If the sample is not conductive, a thin conductive layer can be applied. This is often done using gold.

Reading the Criss-Crossed Lines: TEM

Transmission electron microscopy (TEM) is also used to visualize materials using electrons. Samples must be thin enough that the electrons transmit through them and then onto a detector below the sample. TEM can be used to look at the morphology of materials and also probe the crystalline properties. When crystals are present the electrons are diffracted, similar to what occurs with XRD. The diffracted electrons provide a picture with lines crisscrossing each other, and the lines represent the crystal lattice. The distance between the lines can be measured to determine the d-spacing (see Chapter 18). Not unlike SEM, the sample must be placed in an ultra-high vacuum (UHV) environment because the mean free path of electrons in air is very short.

Characterizing Surface Chemistry: XPS

X-ray photoelectron spectroscopy (XPS) is used to gather data about the chemistry and composition of a material; it's primarily a surface technique. Soft x-rays are incident on a material, the electrons of the atoms are excited, and some are ejected from the electron shell. The electron that's ejected is called a photoelectron, and it has a specific energy depending on the atom and the shell. This is used to work out what atoms are present in the material.

Evaporating Materials: TGA

Thermogravimetric analysis (TGA) is used to determine the decomposition temperature of a material. A material is placed in a small pan and then heated under a gaseous environment. When it evaporates, the pan becomes lighter due to the evaporation of the material as it decomposes. The temperature at which this occurs helps determine the quality of a material because weaker bonds break apart and evaporate sooner than more strongly bonded materials. It can be used to determine the quality of crystal materials, and it can be used to determine the mass loss under specific experimental conditions.

Cyclic Voltammetry

Cyclic voltammetry (CV) can be used to determine if a material will undergo a redox reaction, and test whether the reaction is reversible (cyclic). A material is tested using an applied voltage, the voltage source supplies electrons, so it can be used to test the oxidation and reduction properties of a material. Then the current potential can be reversed, and the material can be tested again to measure what potential is required for the reverse reaction to occur. Because it can test the dynamics of electron transfer reactions, it can be applied to understand catalytic reactions, to analyze stoichiometry of complex compounds, and can determine the bandgap of photovoltaic materials.

Tracking Electron Spin: EPR

Electron paramagnetic resonance (EPR), also called electron spin resonance (ESR) spectroscopy, is used to determine the electron spin for atoms with unpaired electrons. Most atoms have electrons that are paired together, because this makes them more stable. But radicals do not have spin pairs, and they can be tested using this method. Free radicals are often short-lived species, but they are also very important in the outcome of a chemical reaction. So when you can track their progress, you can better understand the mechanisms of chemical reactions.

Chapter 22

Ten Experiments

*T*o do these experiments, you should always wear some form of eye protection so you can protect your eye sight in the event of any unwanted spills or reactions. You should also wear closed-toed shoes so that spillages do not hurt your feet and toes. You should always start with small amounts when mixing chemicals to avoid any large mishaps. All of these experiments can be carried out at home, or they could be used as a chemistry demonstration in a classroom.

Using some common household products and practicing some kitchen chemistry, you can perform some pretty interesting inorganic chemistry experiments. In this chapter, we describe ten easy-to-perform inorganic chemistry experiments.

Turning Blue: The Clock Reaction

There's a chemistry battle between starch and vitamin C. The starch wants to turn blue as it's reduced, but the vitamin C keeps it from being reduced and turning blue. Eventually, an iodine starch complex forms and gives the solution a blue color. In this experiment, you change the amount of time it takes for the liquids to change color.

Start by crushing up a vitamin C tablet into a powder. Add it to 2 ounces (60 mL) of warm water and stir it for 30 seconds. Transfer 1 teaspoon (5 mL) of the vitamin and water mixture into a new cup and then add 2 ounces (60 mL) of warm water + 1 teaspoon (5 mL) of iodine (tincture of iodine 2% concentration). This should make the iodine solution turn clear. In another clean glass, make a new solution of 1 tablespoon (15 mL) hydrogen peroxide (3% concentration) plus ½ teaspoon (2.5 mL) liquid starch dissolved in 2 ounces of warm water. Pour the last two solutions together, and mix them by pouring back and forth between the glasses a few times. The reaction should occur shortly afterwards as the liquid that was being poured between the two cups will suddenly turn blue.

Forming Carbon Dioxide

Put a teaspoon of baking soda into a glass jar and then place a lit candle on top. Add some water to the baking soda, but be careful not to put out the flame in the candle. As the water reacts with the baking soda, it produces carbon dioxide and this can't combust (not without a catalyst present). Increasing concentrations of carbon dioxide starves the flame of oxygen. As the remaining oxygen is burned up by the candle, it's essentially being used to oxidize the carbon from the wax. When the oxygen is bonded to the carbon, as either CO or CO_2 it can't feed the candle flame, and the candle goes out.

The Presence of Carbon Dioxide

In the presence of lime water, carbon dioxide forms calcium carbonate. This is insoluble in water, so it precipitates out of solution. To test this, breathe through a straw into a solution of lime water, like blowing bubbles. This works according to the following formula:

$$Ca(OH)_2 \text{ (aq)} + CO_2 \text{ (g)} \rightarrow CaCO_3 \text{ (s)} + H_2O \text{ (l)}$$

When all the lime has reacted, the product is calcium carbonate, which can be concentrated by evaporating the water.

Mimicking Solubility

Colloidal stability is a temporary phenomenon associated with insoluble substances of a particularly small size, often with nanometer dimensions. The stability comes from the small size and the interaction between the electrical charges of the colloid and the solvent. This interaction gives the tiny colloid some temporary stability that mimics solubility (see Chapters 8 and 19).

Start by crushing the tip of a lead pencil using a mortar and pestle in the presence of a tannin (can try using wine, it has tannins). It can take up to 15 minutes to pulverize the pencil tip enough. Then add water. A dark, opaque liquid forms and should be poured off and diluted so that three parts are water and one part is your test solution. The colloidal tannin solution forms a dark opaque solution that is so soluble it can even pass through filter paper. A solution made without tannin doesn't form an even-looking solution because the graphite isn't colloidal. After a few days, the colloids settle to the bottom, indicating that the colloids are stable only for a certain period of time, and they are not a true solution.

Separating Water into Gas

This experiment uses electricity to split water molecules into hydrogen and oxygen components. Fill up a tray or basin with tap water to make a water bath; then set up a pneumatic trough by placing electrode terminals into two upside-down beakers (each filled with water) in the water bath.

The opposing charges of the atoms in the water molecule are attracted to the positive and negative electrode terminals. As this happens, the top of the inverted glass fills with gas and starts to displace the water downwards. Hydrogen forms at the negative pole, and oxygen forms at the positive pole.

The best electrodes to use are made of platinum, but this experiment can be done with copper, for example. You can connect five-volt square batteries in series to increase the current, which in turn speeds up the reaction.

Testing Conductivity of Electrolyte Solutions

Water by itself isn't a good conductor of electricity because it has only a small amount of electrolytes present. (Electrolytes are charged ions that can help transmit current; see Chapter 8 for details.) The more electrolytes in a solution, the more conductivity the solution displays.

Hook up a battery to a light bulb by connecting the negative terminal to the bulb, connect the positive terminal to the water, and have another piece of wire going to the other terminal of the light bulb. This should be a closed circuit whereby the concentration of electrolytes in the solution alters the current allowed to pass through the solution. The amount of current can be detected by noting the brightness of the light bulb. Add electrolytes from something like coconut water, or brine, to the glass of water, and observe: The light bulb should start to light up more, illustrating that electricity is being conducted by the electrolytes in solution.

You can try different solutions and concentrations to see what difference it makes to the conductivity. Just look to see how bright the light bulb is glowing.

Lemon Batteries

A lemon is full of electrolytes and generates charge between two metals that have different oxidation states. At opposite ends of the lemon, place a galvanized nail (with zinc) and a piece of copper. The galvanized nail (with zinc) acts as a negative terminal, and the copper piece acts a positive terminal. The different redox potential of the zinc and the copper creates a flow of electrons toward the copper. If you hook up four lemon batteries in series, you should create enough current to light a small light-emitting diode (LED) light bulb.

Purifying Hydrogen

React metal with a small amount (5 mL) of sulfuric acid. This forms a metal sulfate, and hydrogen is liberated from the sulfuric acid. Best done with zinc metal found in a penny coin. Hydrogen rises and can be collected by an inverted piece of glassware. When pouring hydrogen between containers you have to remember to pour it upwards. Because it's so light, it tries to float up. The presence of hydrogen can be tested by placing a lit match close to it to cause the hydrogen to ignite and make a sound like a squeaky pop.

Colorful Flames

This experiment requires the use of a household product with colorant, such as bleaching powder with calcium chloride, which makes an orange flame; sodium chloride (table salt), which makes a yellow flame; borax, which makes a yellow/green flame; or magnesium sulfate from Epsom salts, which makes a white flame.

Place one of these colorants into water, and soak some sawdust or a pine-cone in the solution for several hours so that it gets absorbed into the wood fibers. Remove the pinecone or sawdust from the solution, and let it dry completely. When the material is dry, it can be burned and produces a flame color depending on the mineral that is used for a colorant.

Making Gunpowder

Gunpowder is a compound made by mixing potassium chloride, KCl, found in low-fat salt, with nitrate from ammonium nitrate, NH_4NO_3, found in the cold pack from a first aid kit.

Add NH_4NO_3 to about 100 mL of water and stir it until it dissolves completely. Pour the solution through a coffee filter into glass container that has potassium chloride (KCl) at the bottom. Gently heat the mixture, and stir the mixture until all the KCl is dissolved, taking care not to tap the glass too much so that you crack it. Then put the container in the freezer for two hours. (The cold speeds up crystallization.) Crystals of potassium nitrate with the consistency of slush form at the bottom. Pour off the excess liquid to continue drying them.

When dry, in a nonmetallic container so that no sparks form, crush the crystals into a fine powder. (The chance of igniting the saltpeter crystals is quite low, but it's best to be safe and use a nonmetal container.) It works by the following equation:

$$NH_4NO_3 + KCl^- \rightarrow KNO_3 + NH_4Cl$$

Take the crushed, dried crystals, and add them to an equal amount of table sugar to form a makeshift gunpowder.

Chapter 23

Ten Inorganic Household Products

*I*norganic chemicals are commonly used for household cleaning and cooking. This chapter describes a few of the inorganic materials you are likely to find under your kitchen sink or in your laundry room, for example.

Salting Your Food

Every household has salt. It's an important compound that's widely available. The word *salary* harks back to the days when people were paid in salt. It's very soluble in water and is used to draw flavors out of foods so food tastes more prominent. It's also used to dry and preserve foods so that they can be stored for long durations (and this was very important prior to the use of refrigeration).

Salt readily absorbs moisture, so some people put pieces of rice inside their saltshaker because the rice competes for water vapor, thus keeping the salt dry and free flowing.

Salt is produced by three methods. Large deposits can exist underground and can be mined using conventional shaft mines. Or water can be passed through the salt deposits, causing it to flow out from the ground where it is then crystallized. Or sea water can be dried in the sun, leaving behind salt beds. This produces rock salt, evaporated salt, and sea salt, respectively.

Bubbling with Hydrogen Peroxide

Hydrogen peroxide (H_2O_2) is a colorless effervescent solution. It's a common product used to clean wounds and sores. When added to water (or any moist environment), it immediately releases oxygen. This creates bubbles that break apart dirt and foreign material in a wound and also oxidizes (and kills) bacteria. The reaction can be quite vigorous and for this reason it's sold in stores in very dilute concentrations of around 4%.

It can used as a cleaning agent to clean stains from clothing, and also as an agent to clean burned-on food in cooking pans. You know the brown stuff at the bottom of the pan that used to be your baked beans? If you gently heat H_2O_2 in the pan, it oxidizes the carbon (from the burned stuff), causing it to break up and be easily be washed away. Then you can have a nice clean pan again.

Be careful when doing this. Make sure to let the steam evaporate — don't let it build up pressure!

Concentrated solutions can be dangerous because of the vigorous liberation of oxygen. If it's not stored correctly, it can cause a container to rupture due to the buildup of gas. Also, if it's spilled on your body, it can cause severe burns to your skin. And, if it's spilled on flammable materials such as wood, oils, or other organic materials, it can cause a fire to start, and this would be exasperated by the oxygen being released that would fuel the fire more.

Baking with Bicarbonate

Baking soda that you buy at the store is the compound sodium bicarbonate, ($NaHCO_3$). It has many household uses — too many to list here, but here's a sampler to get you started.

- ✔ **Baking:** Baking soda is added to recipes to cause breads and other dough to rise. The raising of the dough comes from the release of CO_2 gas during baking.
- ✔ **Antacid:** Baking soda is an alkaline (basic) material and can ease heartburn by neutralizing stomach acids.
- ✔ **Odor control:** Some people put boxes of baking soda in their kitchens or refrigerators to control food odors. Sodium bicarbonate is a very active material and absorbs and neutralizes the molecules that smell bad.

✓ **Stain removal:** Baking soda can be used to remove a whole slew of difficult stains and odors because of how alkaline or basic it is. For example,
soak stained material with a solution of baking soda in water, then rinse
with water and place in the wash. It can clean blood stains, sweat stains,
fruit and wine stains, and even vomit stains (including the smell). It can
also be used on carpets and many other places around the home such
as in the kitchen, bathroom, living room, and even outdoors in the yard
to clean oil stains or a greasy grill, for example.

Whitening with Bleach

The term *bleach* refers to a number of different materials that have the effect
of removing color from fabrics. Typically, the bleach you buy in a shop is
chlorine-based and contains sodium hypochlorite (NaClO).

Bleach can be used for cleaning all around your house, especially places that
require extra disinfecting, like in the toilet bowl.

Some bleaches function as a strong oxidizing agent, thus it oxidizes the *chromophores,* or color molecules, in your clothes and turns them white. Other
bleaches are strong reducing agents, and they break the bonds in the chro-
mophore, which also takes the color out of the clothing.

Keep in mind that bleach is a toxic material and should be handled with care.
The same strong oxidizing and reducing abilities that kill bacteria and turn
your clothing white can also damage skin tissue both externally and internally.

Using Ammonia in Many Ways

Household ammonia is typically a mixture of 10 percent pure NH_3 and water.
Concentrated ammonia is NH_3, and this is the kind of stuff you might find
farmers using as a fertilizer. It reacts quickly in the presence of water to form
ammonium hydroxide (NH_4OH).

Ammonia has a very strong scent and can burn the inside of your nose, thus
damaging the mucus membrane. Be careful when using this material. Ever
notice the sharp smell from a cat litter box that makes your eyes tear? That
comes from dilute concentrations of ammonia that are present in urine.

Similar to bleach, ammonia can whiten the color of materials. Ammonia can be gathered by concentrating urine; this was used to clean clothing, as well as make tooth paste and mouthwash by the ancient Romans.

Ammonia is also used industrially, particularly in agriculture where it is used as a fertilizer. But don't try this at home. Household ammonia often has additives that make it unsuitable for fertilizing your garden.

Killing Pests with Borax

Borax is commonly used as a pesticide and fungicide. It's made of sodium borate ($NaBO_2$), but it's most often found as the hydrated compound, $Na_2B_4O_7 \cdot 10H_2O$. It can also be used as a cleaning agent, or to soften hard water.

Sodium borate is a white powdery material that, when diluted in water, forms a cleaning solution. Powdered borax is used as a deterrent for pests such as mites and cockroaches. It can be caustic in high doses, but in low concentrations is an effective disinfectant. It's commonly found in laundry detergents, some hand soaps, and it can also be found in teeth bleaching solutions.

Soothing Babies with Talc

Talc is the softest mineral known. It's composed of hydrated magnesium silicate, and is commonly used in one of two forms: $H_2Mg_3(SiO_3)_4$ or $Mg_3Si_4O_{10}(OH)_2$.

It's used to soothe diaper rash or other skin irritations that occur from too much moisture, because it absorbs large amounts of water.

Talc is one of the finest powders you'll find in a household. The grain sizes are on the order of a few hundred microns. That makes it just about visible to the naked eye, but only if you have really good eyesight.

Because talc powder is so fine-grained you should be careful to not directly inhale it into your lungs. It can become trapped and cause respiratory problems.

Cleaning with Lye

Lye is the common name given to hydroxides of either sodium or potassium, these are both alkali metals and they can be quite reactive and caustic, they are described in greater detail in Chapter 11. *Soda lye* is made of sodium hydroxide, (NaOH). *Potash lye* is composed of potassium hydroxide, (KOH). Lye is an effective cleanser because of its caustic nature; it's a strong base (see Chapter 5 for details on acid-base chemistry).

There is a long history (thousands of years) of making and using lye. It can be extracted from the ashes of burned out hardwoods using baking soda, or slaked lime and water. The lye can then be used to dissolve animal fats to make soaps. Remember in Chapter 5 the word *alkaline* comes from the Arabic word *to roast,* which originates from the roasting of ashes to make soaps.

Concentrated lye can be caustic to humans, and it can burn skin if not rinsed adequately with water after contact — otherwise, you run the risk of making soap out of your flesh!

Scratching Stainless Steel

Steel is a metal alloy that's made by mixing a small amount of carbon into iron. This increases the strength of the metal, so you can make lighter and stronger tools or gadgets.

Stainless steel is made of the same basic material, but it has another metal added to it, usually chromium. Chromium is added because it's nearly the same size as iron , so they pack neatly together forming strong bonds with the iron atoms in the solid. Another metal that might be used is titanium, because it's also stable and strong. Both chromium and titanium oxidize to form a protective layer that prevents the underlying metal from rusting.

The way these metals form a protective oxide, or *rust layer,* is the unique property of stainless steel that makes it so useful as a material for cooking tools and eating utensils. When the oxide layer is scratched away and the metal is exposed, it causes the underlying steel to oxidize again. This oxide layer forms quickly and is quite strong. The metal inside is protected by the outer oxide layers.

Stainless steel is a material that rusts very fast, but the rust is strong and ends up protecting the material.

Wrapping It Up with Aluminum Foil

Aluminum is the wonder material of the 20th century. In the kitchen it's found as aluminum foil, a lightweight, flexible sheet of metallic aluminum. It's usually covered by a thin layer of grease to help prevent oxidation. Aluminum is resistant to breakdown over a wide range of conditions, so you can put it in the freezer or the oven, wrap your sandwiches with it, and not be afraid that the aluminum is dissolving into your food.

You should avoid placing it in a microwave oven. It heats up very fast due to its thinness, and in a worst-case scenario could start a fire. This is especially true if there are any kinks in the foil, because the sharp edges from the crinkles heat up very fast.

Aluminum takes large amounts of electrical energy to create, so it's wise to recycle as much as possible.

Glossary

achiral: No handedness, a molecule or object that has no special handedness. In other words, the mirror image is superimposable on the original object.

activation energy: The energy required to transform a chemical compound from a ground state to a transition state during a chemical reaction. Most chemical reactions have activation energies that range from 40-100 kJ/mol.

addition reactions: When two atoms mix together and bind with no parts left out; in short, the combination of two or more molecules to form a larger compound.

adducts: The name given to the product of an addition reaction.

Born-Haber cycle: An important thermodynamic calculation that is used to quantify the energy involved in making ionic compounds.

Charge density: Symbol ρ it's a measure of how much, and how tightly packed the charge is around an atom. It can be calculated by dividing the amount of charge (z), with the radius (r); therefore, the charge density is equal to z/r. According to Fajan's rule, the charge density of a cation affects the extent of covalent bonding.

chiral: To have handedness, molecules are nonsuperimposable on their own mirror image. There is a left-handed and a right-handed form for chiral molecules. The opposite is to be achiral.

chiral center: The place in the molecule where all the symmetry planes coincide. This is usually the position of the central atom in a complex.

cis-trans isomers: Used to describe the relative positions of ligands with respect to each together. *Cis* means together; *trans* means across from.

complex/compound: Name given to a compound that has a discreet formula, shape, and size.

complexing constant: A measure of the strength between a central atom and bonded ligands.

conformation: A way to describe what the chemical looks like in three-dimensional space.

conjugation: Alternating single and double bonds, where there is overlapping p orbitals. Common for molecules with a carbon framework.

constitutional isomers: Isomers that are related by the same chemical formula, but the atoms are connected in different orders.

coordination numbers: The number of ligands that are bonded to a metal define the coordination number; the value indicates the number of bonds the metal is involved.

covalent bond: The sharing of electrons between to atoms, covalent bonds are typically quite strong. They can be single, double, or triple bonds. Covalent bonds have directionality.

d,l form: Notation used to define the rotation direction of polarized light from interaction with an isomer.

dehydration: Reaction that when carried out consumes H_2O molecules. Water is lost from the starting materials.

delocalization: Describes how electrons spread out across a material when they are not tied to the atom from which they originated. Common for conjugated pi electron systems and in bulk metals.

dipole: To have two poles, like a north pole and a south pole. When the distribution of charges in a molecule is uneven, one side is more positively charged than the other, and the molecule is said to have a dipole.

dipole moment: Symbol μ. This is used to describe the position of electron density relative to the center of mass of the atom or molecule; this gives charge to the atom or molecule.

electrophile: A species that accepts an electron pair from a nucleophile.

enantiomers: Chiral molecules that have a mirror-image relationship; they are stereoisomers.

endothermic: When a reaction takes place, energy is required to make it come to completion.

energy: A physical quantity that's a measure of how much work is required to complete a reaction. Measured in Joules, J.

enthalpy: Symbol H, it's a thermodynamic state function used to calculate the energy requirement of chemical reactions at constant pressure. When the enthalpy change is negative, the reaction is exothermic; when the value is positive, it's an endothermic reaction. ΔH is a measure of the amount of heat absorbed or produced during a chemical reaction.

entropy: Symbol S, this is an important thermodynamic state function that measures the degree of randomness or disorder for a given system or chemical reaction. It measures the capacity for a reaction to occur according to thermodynamics. ΔS is a measure of the change of disorder in a system.

exothermic: A reaction is exothermic when a reaction takes place and heat is liberated.

Fajan's rule: Used to determine qualitatively the degree of covalent bonding in ionic materials. Cations experience higher covalent character according to increasing charge density. In anions, the degree of covalent bonding increases with increasing polarizability of the ion.

Fermi energy: Symbol E_F. This is a measure of energy that a material has at absolute zero temperature.

Fermi level: A measure of the highest occupied energy level of a molecule; it's different in value from the Fermi energy because most reactions are not carried out at absolute zero temperature.

fluxionality: The capacity for molecules to rearrange themselves, often used in terms of ligand rearrangements around a metal center for a given coordination complex.

frontier orbitals: The orbitals that are responsible for bonding between atoms and molecules, the orbitals most overlap in space but also be similar in energy. In a nucleophile, they're the highest occupied molecular orbital (HOMO). In electrophiles, they're the lowest unoccupied molecular orbital (LUMO).

geometric isomers: Isomers that have the same connections among the atoms but have different spatial orientations among the metal-ligand bonds.

Gibbs free energy: Symbol G, also referred to as the Gibbs energy, or the Gibbs function is an important concept in thermodynamics. It's used to measure when thermodynamic equilibrium is reached for given reaction within a closed system. It's related to the enthalpy change (H), entropy (S), and temperature (T):

$$G = H - TS, \text{ and } \Delta G = \Delta H - T\Delta S$$

He_2^+ is the capacity for a system to perform work, when $He_2^+ = 0$ the reaction is at equilibrium, when $He_2^+ < 0$ the reaction is spontaneous, when $He_2^+ > 0$ the reaction requires energy.

half-life: Written at $T^{1/2}$, it represent the time it takes for a radioisotope to decrease to half of its original concentration.

Hess's law: For any given chemical reaction, the total energy required to complete the reaction is the same regardless of the number of steps required to complete the reaction. Forms the basic understanding required to complete a calculation using the Born-Haber cycle.

Hund's rule: Used as a guideline to determine the lowest energy state of an atom.

hybrid: Used to describe the mixing of molecular orbitals (s, p, d, f) such that they hybridize to form equivalent lone pairs and bonding pairs orbitals. s-p hybridized orbitals are common for many organic carbon compounds.

hydration: When water is added to a molecule in a chemical reaction. Can be covalently bound or be bonded according to noncovalent interactions, such as in the case of the hydration sphere.

intermediate: Chemical reactions may occur in several steps, an intermediate describes a product that is formed prior during one of the steps. Different from a *transition state*.

ionic bond: An ionic bond is formed when there is a transfer of electrons from one atom to another atom. It's based on electrostatic or columbic interactions.

ions: A species that carries an electric charge.

isomers: Used to describe a series of compounds that have the same molecular formula but are different in shape due to difference in bonding arrangements.

joule: The SI unit for energy, symbol J.

kinetics: A study of the rate of a reaction. It's affected by conditions such as temperature, concentration, and the use of a catalyst.

Lewis structure: Structural representation of how the valence electrons are located about a molecule or atom, it's used to show the presence of bonding pairs and nonbonding lone pairs.

lone-pair electrons: Also called nonbonding electron pairs. This is a pair of electrons that don't get shared with another atom or molecule.

mechanism: The complete path by which a chemical reaction takes places, including all the intermediate steps and transition states that are formed before the reaction has come to completion.

mischmetal: The name given to a metal compound that contains two metals whereby one of the metals is a rare Earth metal.

miscibility: Used to describe the capacity for two liquids to form a single solution. For example, polar liquids dissolve in other polar liquids and are said to be miscible, but a nonpolar liquid (oil) and a polar liquid (water) do not mix. They are immiscible, and they will form into two separate layers.

molarity: Symbol M; the measure of concentration of chemical solutions, the number of moles of a chemical dissolved into liter of solvent: moles/liter.

mole: Symbol mol; the SI unit for an amount of a substance; it's compared to the number of atoms in 0.012 kg of carbon-12.

molecular orbital theory: Approach used to describe the way in which atomic orbitals interact when atoms combine and form a molecule.

nucleophile: In organic chemistry it's a species that donates the electron pair; synonymous with a Lewis base.

opaque: A material is opaque when it is not see-through, or it doesn't allow light to pass through.

oxidation state: Also known as the oxidation number, a measure for how many electrons an atom needs to become a neutral species. Electrons can be added, or removed to reach the neutral atom. In a chemical reaction if the oxidation number increases, the element is reduced.

periodic table: A table of the elements that are ordered according to a successive increase in the atomic number.

pi (π) bond: Bonding between two species in which there is a sideways orbital overlap, leading to a situation that looks like an atomic p-orbital.

polarizability: Used to describe how easy it is for an atom of molecule to change and distort the electron density of the orbital shell around nucleus.

racemic mixture: When a solution contains quantities of both left- and right-handed stereoisomers.

redox: Short for reduction/oxidation reactions; in redox reactions, electrons are exchanged in the reaction, which cause the oxidation numbers to change for the atoms that are involved; a reaction that includes a species that's reduced (gain of electrons) and another that's oxidized (loss of electrons).

resonance effects: Used to describe the situation when a chemical species donates or withdraws electrons through orbital overlap with neighboring pi bonds. In resonance processes, electrons are delocalized and the energy of the electrons are stabilized.

sigma (σ) bond: A covalent bond that's symmetric about the bonding axis. It looks similar to the atomic s orbital, thus the reason for the name sigma, for s. Most covalent bonds are sigma in character. It formed due to the head-on interaction of covalent bonds.

solubility: A measure of how much mass can be added to a liquid that will form a stable solution. It's used to measure how much of a solute can be added to a solvent and still be a stable solution. If too much solute is added, it's said to be a supersaturated solution. Solubility is affected by temperature.

solubility product: Symbol K_{sp}; defines the equilibrium between a salt and its aqueous ions. When the value of K_{sp} is high the solubility is high, when the value of K_{sp} is low the product has a low solubility.

spectroscopy: The study of how electromagnetic radiation interacts with matter.

stereochemistry: Used to describe the way atoms and molecules are arranged in three-dimensional space.

stereoisomers: Isomers that are different from each other according to how they are arranged in physical space.

steric strain: When ligands come into very close proximity to each other, the conformation is strained because of steric repulsion because the space is limited or tight.

stoichiometry: The ratio of atoms in a molecule.

substitution reactions: Where one constituent in a molecule is substituted for another constituent atom, and there are no left-over atoms after the reaction is completed.

symmetry plane: When a molecule is symmetric, the plane that dissects the molecule through the middle is known as the symmetry plane.

thermodynamic control: Reactions that create products with the lowest energy and most stable product are said to be controlled by thermodynamics. Not all reactions create the most thermodynamic product; some are controlled by kinetics instead.

transition state: The highest energy state for a reactant because it changes from reactants to products. Transition states are very short lived and can't be isolated; contrasts with an intermediate state that can be isolated.

valence shell electron pair repulsion theory: The method used to describe the optimal arrangement and geometry of atoms and molecules. It's used to describe the shape of simple molecules.

valency: A measure of the number of electrons in the valence shell of an atom. It highlights the capacity of an atom for undergoing a reaction, but it's used much less often compared to the oxidation number.

van der Waals forces: An intermolecular force between molecules. It's a low-energy, short-range force.

Index

• D •

• *K* •

• *L* •

• *M* •

• T •

• U •

• V •

• W •

• X •

• Z •

Apple & Mac

iPad For Dummies,
5th Edition
978-1-118-49823-1

iPhone 5 For Dummies,
6th Edition
978-1-118-35201-4

MacBook For Dummies,
4th Edition
978-1-118-20920-2

OS X Mountain Lion
For Dummies
978-1-118-39418-2

Blogging & Social Media

Facebook For Dummies,
4th Edition
978-1-118-09562-1

Mom Blogging
For Dummies
978-1-118-03843-7

Pinterest For Dummies
978-1-118-32800-2

WordPress For Dummies,
5th Edition
978-1-118-38318-6

Business

Commodities For Dummies,
2nd Edition
978-1-118-01687-9

Investing For Dummies,
6th Edition
978-0-470-90545-6

Personal Finance
For Dummies,
7th Edition
978-1-118-11785-9

QuickBooks 2013
For Dummies
978-1-118-35641-8

Small Business Marketing Kit
For Dummies,
3rd Edition
978-1-118-31183-7

Careers

Job Interviews
For Dummies,
4th Edition
978-1-118-11290-8

Job Searching with
Social Media
For Dummies
978-0-470-93072-4

Personal Branding
For Dummies
978-1-118-11792-7

Resumes For Dummies,
6th Edition
978-0-470-87361-8

Success as a Mediator
For Dummies
978-1-118-07862-4

Diet & Nutrition

Belly Fat Diet For Dummies
978-1-118-34585-6

Eating Clean For Dummies
978-1-118-00013-7

Nutrition For Dummies,
5th Edition
978-0-470-93231-5

Digital Photography

Digital Photography
For Dummies,
7th Edition
978-1-118-09203-3

Digital SLR Cameras &
Photography For Dummies,
4th Edition
978-1-118-14489-3

Photoshop Elements 11
For Dummies
978-1-118-40821-6

Gardening

Herb Gardening
For Dummies,
2nd Edition
978-0-470-61778-6

Vegetable Gardening
For Dummies,
2nd Edition
978-0-470-49870-5

Health

Anti-Inflammation Diet
For Dummies
978-1-118-02381-5

Diabetes For Dummies,
3rd Edition
978-0-470-27086-8

Living Paleo For Dummies
978-1-118-29405-5

Hobbies

Beekeeping
For Dummies
978-0-470-43065-1

eBay For Dummies,
7th Edition
978-1-118-09806-6

Raising Chickens
For Dummies
978-0-470-46544-8

Wine For Dummies,
5th Edition
978-1-118-28872-6

Writing Young Adult Fiction
For Dummies
978-0-470-94954-2

Language &
Foreign Language

500 Spanish Verbs
For Dummies
978-1-118-02382-2

English Grammar
For Dummies,
2nd Edition
978-0-470-54664-2

French All-in One
For Dummies
978-1-118-22815-9

German Essentials
For Dummies
978-1-118-18422-6

Italian For Dummies
2nd Edition
978-1-118-00465-4

 **Available in print and e-book formats.**

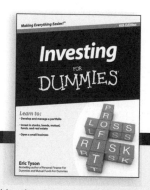

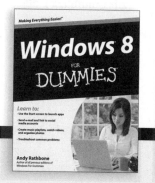

Math & Science

Algebra I For Dummies,
2nd Edition
978-0-470-55964-2

Anatomy and Physiology
For Dummies,
2nd Edition
978-0-470-92326-9

Astronomy For Dummies,
3rd Edition
978-1-118-37697-3

Biology For Dummies,
2nd Edition
978-0-470-59875-7

Chemistry For Dummies,
2nd Edition
978-1-1180-0730-3

Pre-Algebra Essentials
For Dummies
978-0-470-61838-7

Microsoft Office

Excel 2013 For Dummies
978-1-118-51012-4

Office 2013 All-in-One
For Dummies
978-1-118-51636-2

PowerPoint 2013
For Dummies
978-1-118-50253-2

Word 2013 For Dummies
978-1-118-49123-2

Music

Blues Harmonica
For Dummies
978-1-118-25269-7

Guitar For Dummies,
3rd Edition
978-1-118-11554-1

iPod & iTunes
For Dummies,
10th Edition
978-1-118-50864-0

Programming

Android Application
Development For
Dummies, 2nd Edition
978-1-118-38710-8

iOS 6 Application
Development For Dummies
978-1-118-50880-0

Java For Dummies,
5th Edition
978-0-470-37173-2

Religion & Inspiration

The Bible For Dummies
978-0-7645-5296-0

Buddhism For Dummies,
2nd Edition
978-1-118-02379-2

Catholicism For Dummies,
2nd Edition
978-1-118-07778-8

Self-Help & Relationships

Bipolar Disorder
For Dummies,
2nd Edition
978-1-118-33882-7

Meditation For Dummies,
3rd Edition
978-1-118-29144-3

Seniors

Computers For Seniors
For Dummies,
3rd Edition
978-1-118-11553-4

iPad For Seniors
For Dummies,
5th Edition
978-1-118-49708-1

Social Security
For Dummies
978-1-118-20573-0

Smartphones & Tablets

Android Phones
For Dummies
978-1-118-16952-0

Kindle Fire HD
For Dummies
978-1-118-42223-6

NOOK HD For Dummies,
Portable Edition
978-1-118-39498-4

Surface For Dummies
978-1-118-49634-3

Test Prep

ACT For Dummies,
5th Edition
978-1-118-01259-8

ASVAB For Dummies,
3rd Edition
978-0-470-63760-9

GRE For Dummies,
7th Edition
978-0-470-88921-3

Officer Candidate Tests,
For Dummies
978-0-470-59876-4

Physician's Assistant Exam
For Dummies
978-1-118-11556-5

Series 7 Exam
For Dummies
978-0-470-09932-2

Windows 8

Windows 8 For Dummies
978-1-118-13461-0

Windows 8 For Dummies,
Book + DVD Bundle
978-1-118-27167-4

Windows 8 All-in-One
For Dummies
978-1-118-11920-4

 **Available in print and e-book formats.**

Available wherever books are sold. For more information or to order direct: U.S. customers visit www.Dummies.com or call 1-877-762-2974.
U.K. customers visit www.Wileyeurope.com or call (0) 1243 843291. Canadian customers visit www.Wiley.ca or call 1-800-567-4797.

Connect with us online at www.facebook.com/fordummies or @fordummies

Take Dummies with you everywhere you go!

Whether you're excited about e-books, want more from the web, must have your mobile apps, or swept up in social media, Dummies makes everything easier .

Visit Us

Like Us

Follow Us

Watch Us

Join Us

Pin Us

Circle Us

Shop Us

Dummies products make life easier!

- DIY
- Consumer Electronics
- Crafts

- Software
- Cookware
- Hobbies

- Videos
- Music
- Games
- and More!

For more information, go to **Dummies.com**® and search the store by category.

FOR
DUMMIES
A Wiley Brand